Experiments Manual

to accompany

Electronic Principles

Eighth Edition

Albert P. Malvino

David J. Bates

Patrick E. Hoppe
Gateway Technical College

EXPERIMENTS MANUAL TO ACCOMPANY ELECTRONIC PRINCIPLES, EIGHTH EDITION

Published by McGraw-Hill Education, 2 Penn Plaza, New York, NY 10121. Copyright 2016 by McGraw-Hill Education. All rights reserved. Printed in the United States of America. Previous edition © 2007. No part of this publication may be reproduced or distributed in any form or by any means, or stored in a database or retrieval system, without the prior written consent of McGraw-Hill Education, including, but not limited to, in any network or other electronic storage or transmission, or broadcast for distance learning.

Some ancillaries, including electronic and print components, may not be available to customers outside the United States.

This book is printed on acid-free paper.

4 5 6 7 8 9 QVS/QVS 21 20 19 18 17

ISBN 978-1-259-20011-3
MHID 1-259-20011-6

Senior Vice President, Products & Markets: *Kurt L. Strand*
Vice President, General Manager, Products & Markets: *Marty Lange*
Director of Digital Content: *Thomas Scaife, Ph.D*
Vice President, Content Design & Delivery: *Kimberly Meriwether David*
Managing Director: *Thomas Timp*
Global Publisher: *Raghu Srinivasan*
Director, Product Development: *Rose Koos*
Product Developer: *Vincent Bradshaw*
Marketing Manager: *Nick McFadden*
Director, Content Design & Delivery: *Linda Avenarius*
Executive Program Manager: *Faye M. Herrig*
Content Project Managers: *Jessica Portz, Tammy Juran, & Sandra Schnee*
Buyer: *Susan K. Culbertson*
Design: *Studio Montage, St. Louis, MO*
Content Licensing Specialist: *DeAnna Dausener*
Cover Image: *Royalty-Free/CORBIS*
Compositor: *MPS Limited*
Typeface: *10/12 Times New Roman*
Printer: *Quad-Versailles Digital*

Contents

Preface

The 8th edition of *Experiments for Electronic Principles* has been modified to give both the student and the instructor maximum flexibility in performing the experiments while incorporating additional troubleshooting opportunities. This laboratory manual contains 67 experiments to demonstrate the theory in *Electronic Principles*. Sixty-one of the experiments focus on specific topics covered in the textbook. These 61 experiments help the student better understand the individual topics presented throughout the textbook. The additional six System Applications are system-based experiments that tie together related topics covered within the textbook. The early applications tie together related topics using single-stage systems. The later applications present the student with the opportunity to work with multiple-stage systems. The system application-based experiments stand on their own and may be completed without doing the preceding experiments. However, they do tie together the preceding system applications. Each experiment begins with a short discussion. Then, under Required Reading, the entry provides the sections of the textbook that should be read before attempting to do the experiment. In the Procedure section, one or more circuits will be built and tested. Troubleshooting and Critical Thinking sections explore more advanced areas. Questions at the end of each experiment are a final test of what was learned during the experiment.

Optional sections are included in many experiments. They are to be used at the discretion of the instructor. Applications sections demonstrate how to use transducers such as buzzers, LEDs, microphones, motors, photoresistors, phototransistors, and speakers. Additional Work sections include advanced experiments.

The Instructor Resources section of McGraw-Hill's Connect has Multisim-constructed circuits available for download for each of the experiments. Multisim provides the students with an opportunity to gain valuable troubleshooting experience without the potential to damage test equipment or electronic parts. At the end of each experiment, the data tables include columns for data obtained from calculations, measurements using Multisim, and from hands-on measurements. Instructors can assign students to first perform the required circuit calculations, then test the circuit using simulation software, and then build the circuit using actual components. As alternatives, the students can take circuit measurements using only actual components or only circuit simulation. This enables the instructor to offer the laboratory portion of the course using either online or blended formats.

Students will find the experiments in this lab manual to be instructive and interesting. The experiments verify and expand the theory presented in *Electronic Principles*. They make it come alive. When the experiments have been completed, students will have that rounded grasp of theory that can come only from practical experimentation.

Albert P. Malvino
David J. Bates
Patrick E. Hoppe

*To access the Instructor Resources through Connect, contact your McGraw-Hill Representative to obtain a password. If you do not know your representative, go to www.mhhe.com/rep, to find your representative.

Once you have your password, log in at connect.mheducation.com. Click on the course for which you are using *Electronic Principles*. If you have not added a course, click "Add Course," and select "Engineering Technology" from the drop-down menu. Select *Electronic Principles*, 8e and click "Next." Once you have added the course, click on the "Library" link, and then click "Instructor Resources."

Voltage and Current Sources

An ideal or perfect voltage source produces an output voltage that is independent of the load resistance. A real voltage source, however, has a small internal resistance that produces an *IR* drop. As long as this internal resistance is much smaller than the load resistance, almost all the source voltage appears across the load. A stiff voltage source is one whose internal resistance is less than 1/100 of the load resistance. With a stiff voltage source, at least 99 percent of the source voltage appears across the load resistor.

A current source is different. It produces an output current that is independent of the load resistance. One way to build a current source is to use a source resistance that is much larger than the load resistance. An ideal current source has an infinite source resistance. A real current source has an extremely high source resistance. A stiff current source is one whose internal resistance is at least 100 times greater than the load resistance. With a stiff current source, at least 99 percent of the source current passes through the load resistor.

In this experiment, the conditions necessary to achieve a stiff voltage source and current source will be investigated. This experiment will also provide an opportunity to troubleshoot and design both voltage and current sources.

GOOD TO KNOW

Even new components can be out of tolerance. It is a good habit to measure the value of the resistors before use.

Required Reading

Chapter 1 (Secs. 1-3 and 1-4) of *Electronic Principles,* 8th ed.

Equipment

1   power supply: adjustable to 10 V
6   ½-W resistors: 10 Ω, 47 Ω, 100 Ω, 470 Ω, 1 kΩ,
    10 kΩ
1   DMM (digital multimeter)

Procedure

VOLTAGE SOURCES

1. The circuit left of the *AB* terminals in Fig. 1-1 represents a voltage source and its internal resistance R_1.

Before measuring any voltage or current, the approximate value should be known so that the test equipment can be set to the proper range. Examine Fig. 1-1 and estimate and record the load voltage for each value of R_1 listed in Table 1-1. It is important to be able to estimate these rough values and to be able to calculate exact values.

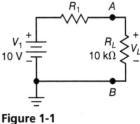

Figure 1-1

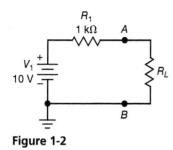

Figure 1-2

2. Sketch the circuit in Fig. 1-1. Measure and record the value of each of the resistors used in this experiment. Build the circuit in Fig. 1-1 using the values of R_1 given in Table 1-1. Measure and adjust the source voltage to 10 V. For each R_1 value, measure and record V_L in Table 1-1.

CURRENT SOURCE

3. The circuit left of the AB terminals in Fig. 1-2 acts like a current source under certain conditions. Estimate and record the load current for each value of load resistance shown in Table 1-2.
4. Sketch the circuit in Fig. 1-2. Build the circuit of Fig. 1-2 using the R_L values given in Table 1-2. Measure and adjust the source voltage to 10 V. For each R_L value, measure and record I_L in Table 1-2.

GOOD TO KNOW

The key to troubleshooting a circuit is to understand how the circuit should work and recognize which component(s) are causing the error through careful measurement and observation.

TROUBLESHOOTING

5. Build the circuit of Fig. 1-1 with an R_1 of 470 Ω. Connect a jumper wire between A and B. Measure the voltage across the load resistor and record V_L in Table 1-3. Record any observations next to Table 1-3.
6. Remove the jumper and open the load resistor by lifting one leg of the resistor from the breadboard. Measure the load voltage between the AB terminals and record in Table 1-3. Record any observations next to Table 1-3.

CRITICAL THINKING

7. Select an internal resistance R_1 for the circuit of Fig. 1-1 to get a stiff voltage source for all load resistances greater than 10 kΩ. Build the circuit of Fig. 1-1 using the selected design value of R_1. Measure the load voltage. Record the value of R_1 and the load voltage in Table 1-4. Record any observations next to Table 1-4.
8. Select an internal resistance R_1 for the circuit of Fig. 1-2 to get a stiff current source for all load resistances less than 100 Ω. Build the circuit with the selected design value of R_1 and a load resistance of 100 Ω. Measure the load current. Record the value of R_1 and the load current in Table 1-4. Record any observations next to Table 1-4.

Experiment 1

Lab Partner(s) _____

PARTS USED	
Nominal Value	Measured Value
10 Ω	
47 Ω	
100 Ω	
470 Ω	
1 kΩ	
10 kΩ	

CALCULATIONS

SCHEMATIC

TABLE 1-1. VOLTAGE SOURCE

R_1	Estimated V_L	Measured V_L	
		Multisim	Actual
0 Ω			
10 Ω			
100 Ω			
470 Ω			

SCHEMATIC

TABLE 1-2. CURRENT SOURCE

R_L	Estimated I_L	Measured I_L	
		Multisim	Actual
0 Ω			
10 Ω			
100 Ω			
470 Ω			

TABLE 1-3. TROUBLESHOOTING

	Measured I_L	
Trouble	Multisim	Actual
Shorted Load		
Open Load		

OBSERVATIONS

TABLE 1-4. CRITICAL THINKING

Source Type	R_1	Measured Quantity	
		Multisim	Actual
Voltage			
Current			

OBSERVATIONS

Questions for Experiment 1

1. The data of Table 1-1 prove that load voltage is: (   )
   (a) perfectly constant;   (b) small;   (c) heavily dependent on load resistance;
   (d) approximately constant.
2. When internal resistance R_1 increases in Fig. 1-1, load voltage: (   )
   (a) increases slightly;   (b) decreases slightly;   (c) stays the same.
3. In Fig. 1-1, the voltage source is stiff when R_1 is less than: (   )
   (a) $0\ \Omega$;   (b) $100\ \Omega$;   (c) $500\ \Omega$;   (d) $1\ k\Omega$.
4. The circuit left of the *AB* terminals in Fig. 1-2 acts approximately like a current   (   )
   source because the current values in Table 1-2:
   (a) increase slightly;   (b) are almost constant;   (c) decrease a great deal;
   d) depend heavily on load resistance.
5. In Fig. 1-2, the circuit acts like a stiff current source as long as the load resistance is: (   )
   (a) less than $10\ \Omega$;   (b) large;   (c) much larger than $1\ k\Omega$;
   (d) greater than $1\ k\Omega$.
6. Briefly explain the difference between a stiff voltage source and a stiff current source.

TROUBLESHOOTING

7. Explain why the load voltage with a shorted load is zero in Table 1-3. Consider using Ohm's
   law to support the explanation.

8. Briefly explain why the load voltage with an open load is approximately equal to the source
   voltage in Table 1-3. Consider using Ohm's law and Kirchhoff's voltage law to support the
   explanation.

CRITICAL THINKING

9. A design requires a current source that must appear stiff to all load resistances less than
   $10\ k\Omega$. What is the minimum internal resistance the source can have? Justify this answer.

10. Optional: Instructor's question.

Thevenin's Theorem

The Thevenin voltage is the voltage that appears across the load terminals when the load resistor is open or removed. The Thevenin voltage is also called the open-circuit or open-load voltage. The Thevenin resistance is the resistance between the load terminals with the load disconnected and all sources are replaced with their ideal internal resistance. The ideal internal resistance for a voltage source is zero ohms, which can be achieved by replacing the voltage source with a jumper wire. The ideal internal resistance for a current source is infinite ohms, which can be achieved by removing the currrent source from the circuit and leaving the two connections open.

In this experiment, the Thevenin voltage and resistance of a circuit will be calculated. The circuit will be built, and these quantities will be measured. Also included are circuit simulation, troubleshooting, and design options.

GOOD TO KNOW

There is a significant difference between replacing the voltage source with a short and shorting out the voltage source. The latter can cause injury and damage the source.

Required Reading

Chapter 1 (Sec. 1-5) of *Electronic Principles,* 8th ed.

Equipment

1 power supply: 15 V (adjustable)
7 ½-W resistors: 470 Ω, two 1 kΩ, two 2.2 kΩ, two 4.7 kΩ
1 potentiometer: 5 kΩ
1 DMM (digital multimeter)

GOOD TO KNOW

Even new components can be out of tolerance. It is a good habit to measure the value of the resistors before use.

Procedure

1. Sketch the circuit in Fig. 2-1*a*, and then calculate the Thevenin voltage V_{TH} and the Thevenin resistance R_{TH}. Record these values in Table 2-1.
2. Sketch the Thevenin equivalent circuit shown in Fig. 2-1*b*. Using the calculated Thevenin values, calculate the load voltage V_L across an R_L of 1 kΩ (see Fig. 2-1*b*). Record V_L in Table 2-2.
3. Also calculate the load voltage V_L for an R_L of 4.7 kΩ as shown in Fig. 2-1*c*. Record the calculated V_L in Table 2-2.
4. Measure and record the value of each of the resistors used in this experiment. Build the circuit of Fig. 2-1*a*, leaving out R_L.
5. Measure and adjust the source voltage of 15 V. Measure V_{TH} and record the value in Table 2-1.
6. Replace the 15-V source with a jumper wire to simulate the ideal internal resistance of the voltage source. Measure the resistance between the *AB* terminals

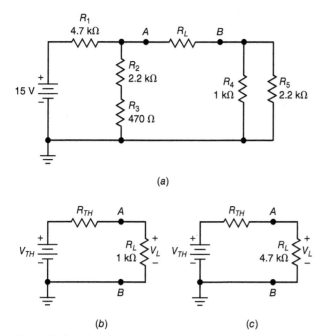

(a)

(b)   (c)

Figure 2-1

using a convenient resistance range of the DMM. Record R_{TH} in Table 2-1. Now replace the wire jumper with the original 15-V source.

7. Connect a load resistance R_L of 1 kΩ between the AB terminals of Fig. 2-1a. Measure and record load voltage V_L (Table 2-2).

8. Change the load resistance from 1 kΩ to 4.7 kΩ. Measure and record the new load voltage.

9. Find R_{TH} by the matched-load method; that is, use the potentiometer as a variable resistance between the AB terminals of Fig. 2-1a. Vary resistance until load voltage drops to half of the measured V_{TH}. Then disconnect the load resistance and measure its resistance with the DMM. This value should agree with R_{TH} found in Step 6.

TROUBLESHOOTING

10. Put a jumper wire across the 2.2-kΩ resistor R_2, the one on the left side of Fig. 2-1a. Estimate the Thevenin voltage and Thevenin resistance for this trouble and record the estimates in Table 2-3. Measure the Thevenin voltage and Thevenin resistance (similar to Steps 5 and 6). Record the measured data in Table 2-3.

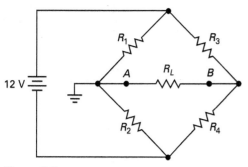

Figure 2-2

11. Remove the jumper wire and open the 2.2-kΩ resistor R_2 of Fig. 2-1a. Estimate and record the Thevenin quantities (Table 2-3). Measure and record the Thevenin quantities.

CRITICAL THINKING

12. Select the resistors for the unbalanced Wheatstone bridge of Fig. 2-2 to meet these specifications: $V_{TH} = 4.35$ V and $R_{TH} = 3$ kΩ. Resistor values must be from those specified under the heading "Equipment." Record the design values in Table 2-4. Build the circuit. Measure and record the Thevenin quantities.

13. Measure the output resistance R_{TH} of the signal generator or function generator that will be used as a signal source for the rest of the experiments in this manual. Use the matched-load method as follows: First, adjust the open-circuit output voltage of the generator to exactly 1 V_{rms} measured by the DMM. Do not change the output voltage adjustment for the balance of this procedure. Then, connect a 1-kΩ potentiometer (connected as a rheostat) as a load resistor on the output of the generator in parallel with the DMM. Now, without changing the amplitude setting of the generator, adjust the potentiometer until the DMM reads exactly 0.5 V_{rms}. Disconnect the DMM and the potentiometer from the generator and measure the resistance of the potentiometer. Record this value in Table 2-5 as the output resistance of the generator. It will be useful in later experiments.

ADDITIONAL WORK (OPTIONAL)

14. Measure the output impedance of a signal generator or a function generator at several frequencies. Draw the Thevenin-equivalent circuit for the generator. Graph z_{out} versus frequency.

15. Use an oscilloscope and a DMM to measure the sinusoidal output voltage of a signal generator or a function generator at various frequencies above 1 kHz. Notice how the measurements differ. One is a visual display, and the other is an rms reading.

Experiment 2

Lab Partner(s) _____

PARTS USED	
Nominal Value	Measured Value
470 Ω	
1 kΩ	
1 kΩ	
2.2 kΩ	
2.2 kΩ	
4.7 kΩ	
4.7 kΩ	

CALCULATIONS

SCHEMATICS

TABLE 2-1. THEVENIN VALUES

	Calculated	Measured	
		Multisim	Actual
R_{TH}			
V_{TH}			

SCHEMATICS

TABLE 2-2. LOAD VOLTAGES

	Calculated V_L	Measured V_L	
		Multisim	Actual
1 kΩ			
4.7 kΩ			

TABLE 2-3. TROUBLESHOOTING

| | Estimated | | Measured | | | |
| | | | Multisim | | Actual | |
	V_{TH}	R_{TH}	V_{TH}	R_{TH}	V_{TH}	R_{TH}
Shorted 2.2 kΩ						
Open 2.2 kΩ						

OBSERVATION

TABLE 2-4. CRITICAL THINKING

| | Measured | | | |
| | Multisim | | Actual | |
	V_{TH}	R_{TH}	V_{TH}	R_{TH}
Design values: $R_1 =$				
$R_2 =$				
$R_3 =$				
$R_4 =$				

OBSERVATION

TABLE 2-5. GENERATOR

Generator output resistance:

$R_{TH} =$ _____

GOOD TO KNOW

The generator R_{TH} and the load resistance form a voltage divider. A stiff voltage source has a load resistance 100 times larger than its R_{TH}.

Questions for Experiment 2

1. In this experiment, the Thevenin voltage was measured with: ( )
   (a) a DMM;   (b) the load disconnected;   (c) the load in the circuit.
2. R_{TH} was initially measured with a: ( )
   (a) voltmeter;   (b) load;   (c) shorted source.
3. R_{TH} was also measured by the matched-load method, which involves: ( )
   (a) an open voltage source;   (b) a load that is open;   (c) varying Thevenin resistance until it matches load resistance;   (d) changing load resistance until load voltage drops to $V_{TH}/2$.
4. Discrepancies between calculated and measured values in Table 2-1 may be caused by: ( )
   (a) instrument error;   (b) resistor tolerance;   (c) human error;   (d) all the foregoing.
5. If a black box puts out a constant voltage for all load resistances, the Thevenin resistance of this box approaches: ( )
   (a) zero;   (b) infinity;   (c) load resistance.
6. Ideally, a voltmeter should have infinite resistance. Explain how a voltmeter with an input resistance of 100 kΩ will introduce a small error in Step 5 of the procedure.

TROUBLESHOOTING

7. Briefly explain why the Thevenin voltage and resistance are both lower when the 2.2-kΩ resistor is shorted.

8. Explain why V_{TH} and R_{TH} are higher when the 2.2-kΩ resistor is open.

CRITICAL THINKING

9. In the manufacturing of automobile batteries, should the batteries have a very low internal resistance or a very high internal resistance? Consider using a voltage divider scenario to support the explanation.

10. Optional: Instructor's question.

Troubleshooting

An open device always has zero current and unknown voltage. The voltage present across the open terminals is determined by inspecting the rest of the circuit. On the other hand, a shorted device always has the zero voltage and unknown current. The current flowing through the shorted device is determined by inspecting the rest of the circuit. In this experiment, component failures will be inserted into a basic circuit and the voltages present across each device will be calculated and then verified through measurement.

GOOD TO KNOW

In the schematic drawing, mark the predicted polarities on the resistors so that the DMM leads can be connected properly.

Required Reading

Chapter 1 (Sec. 1-7) of *Electronic Principles,* 8th ed.

Equipment

1   power supply: 10 V
4   ½-W resistors: 1 kΩ, 2.2 kΩ, 3.9 kΩ, and 4.7 kΩ
1   DMM (digital multimeter)

Procedure

1. Sketch Fig. 3.1 and then calculate the voltage between node *A* and ground. Record the value in Table 3-1 under "Circuit OK."
2. Calculate the voltage between node *B* and ground. Record the value.
3. Measure and record the value of each of the resistors used in this experiment. Connect the circuit shown in Fig. 3-1.
4. Measure the voltages at *A* and *B*. Record these values in Table 3-1.

5. Sketch Fig. 3-1 with R_1 shown as an open (removed). Calculate the voltages at nodes *A* and *B*. Record these values in Table 3-1. Next, open resistor R_1 by removing one leg of the resistor from the circuit. Measure the voltages at nodes *A* and *B*. Record the values in Table 3-1.
6. Repeat Step 5 for each of the remaining resistors listed in Table 3-1.
7. Short-circuit resistor R_1 by placing a jumper wire across it. Calculate and record the voltages in Table 3-1.
8. Repeat Step 7 for each of the remaining resistors in Table 3-1.

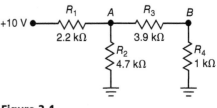

Figure 3-1

Experiment 3

Lab Partner(s) _____

PARTS USED	
Nominal Value	Measured Value
1 kΩ	
2.2 kΩ	
3.9 kΩ	
4.7 kΩ	

SCHEMATIC

CALCULATIONS

TABLE 3-1. TROUBLES AND VOLTAGES

	Calculated		Measured			
			Multisim		Actual	
Trouble	V_A	V_B	V_A	V_B	V_A	V_B
Circuit OK						
R_1 open						
R_2 open						
R_3 open						
R_4 open						
R_1 shorted						
R_2 shorted						
R_3 shorted						
R_4 shorted						

Questions for Experiment 3

1. When R_1 is open in Fig. 3-1, V_A is approximately: ( )
   (a) 0;   (b) 1.06 V;   (c) 1.41 V;   (d) 6.81 V.
2. When R_2 is open in Fig. 3-1, V_B is approximately: ( )
   (a) 0;   (b) 1.06 V;   (c) 1.41 V;   (d) 6.81 V.
3. When R_3 is open in Fig. 3-1, V_A is approximately: ( )
   (a) 0;   (b) 1.06 V;   (c) 1.41 V;   (d) 6.81 V.
4. When R_4 is open in Fig. 3-1, V_B is approximately: ( )
   (a) 0;   (b) 1.06 V;   (c) 1.41 V;   (d) 6.81 V.
5. When R_1 is shorted in Fig. 3-1, V_A is approximately: ( )
   (a) 0;   (b) 2.04 V;   (c) 2.72 V;   (d) 10 V.
6. When R_2 is shorted in Fig. 3-1, V_B is approximately: ( )
   (a) 0;   (b) 2.04 V;   (c) 2.72 V;   (d) 10 V.
7. When R_3 is shorted in Fig. 3-1, V_A is approximately: ( )
   (a) 0;   (b) 2.04 V;   (c) 2.72 V;   (d) 10 V.
8. When R_4 is shorted in Fig. 3-1, V_B is approximately: ( )
   (a) 0;   (b) 1.06 V;   (c) 1.41 V;   (d) 6.81 V.
9. When R_3 is open in Fig. 3-1, V_B is approximately: ( )
   (a) 0;   (b) 1.06 V;   (c) 1.41 V;   (d) 4.92 V.
10. When R_4 is shorted in Fig. 3-1, V_A is approximately: ( )
    (a) 0;   (b) 1.06 V;   (c) 1.41 V;   (d) 4.92 V.

Semiconductor Diodes

An ideal diode acts like a switch that is closed when forward biased and open when reverse biased. One way to test a diode is with an analog ohmmeter. When the diode is forward biased, the ohmmeter is measuring the forward resistance R_F. When the diode is reverse biased, the ohmmeter is measuring the reverse resistance R_R. With silicon diodes, the ratio of R_R to R_F is more than 1000:1.

A diode can also be tested with a digital multimeter (DMM). Most DMMs have a special position (marked with the diode symbol) for testing diodes. To check forward voltage, connect the red lead to the anode (unmarked) and the black lead to the cathode (marked with a colored band). A typical DMM should read 0.5 to 0.7 V for silicon diodes, 0.2 to 0.4 V for germanium diodes, and 1.4 to 2 V for LEDs. A reading near zero indicates a shorted diode, and an overrange indicates an open diode. When the leads are reversed, overrange should be displayed. A reading less than overrange indicates a very leaky diode.

Testing a diode with an ohmmeter or a DMM is an incomplete test because it checks only for major defects such as shorts, opens, or very leaky diodes. The ohmmeter cannot detect more subtle problems. For instance, a slightly leaky silicon diode may have a reverse resistance of only 10 kΩ, low enough to prevent it from working in many circuits. If a slightly leaky diode is tested with a typical DMM, it will pass because the DMM will still indicate overrange when the diode is reverse biased.

The point is this: Testing a diode with an ohmmeter or DMM is conclusive only when the diode fails the test. If it passes the test, the diode may still have some defects that prevent it from working in an ac circuit. Therefore, even though a diode may pass dc testing with an ohmmeter, it must still be checked in a working circuit to verify it's operating as expected.

GOOD TO KNOW

The cathode end of the diode is marked with a colored band.

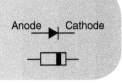

Required Reading

Chapters 2 and 3 (Secs. 3-1 and 3-2) of *Electronic Principles*, 8th ed.

Equipment

1 power supply: adjustable to 10 V
1 VOM (analog multimeter)
1 DMM (digital multimeter)
4 1N4001 diodes
4 ½-W resistors: 1 kΩ, 2.2 kΩ, 4.7 kΩ, and 6.8 kΩ

Procedure

1. In this part of the experiment, the forward and reverse resistance of a diode will be measured and compared against the expected results.

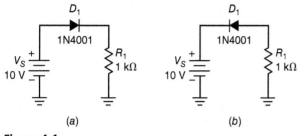

(a)

(b)

Figure 4-1

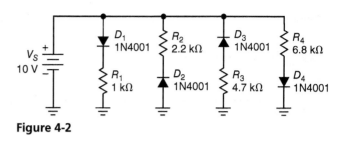

Figure 4-2

2. Should the measured forward resistance of a diode be a low or high value? Record the answer in Table 4-1.

3. Repeat Step 2 for the reverse resistance of a diode.

4. With an analog ohmmeter set to the X100 range, measure the resistance of a 1N4001 in either direction. (The polarity of the red and black leads of the ohmmeter doesn't matter because the red lead may be positive or negative with analog ohmmeters.) Then reverse the leads and measure in the other direction. The meter should display a low reading one way and a high reading the other way. (The exact values don't matter because the resistance of the diode will depend on the ohmmeter range.)

5. Record the readings in Table 4-1 under "Measured 1." If the reverse resistance is too high to read, record "open."

6. Repeat Steps 4 and 5 for two more diodes.

7. Next, use a DMM to test a diode as follows: Select the special diode position (marked with a diode symbol). Connect the red lead to the anode and the black lead to the cathode of a 1N4001. With DMMs, the red and black leads are polarized because the red lead is always positive on the special diode position. The DMM will forward-bias the diode and display the voltage across it. Record the voltage in Table 4-2 under "Measured 1."

8. The reading observed in the preceding step depends on the DMM used. Since a 1N4001 is a silicon diode, the reading should be from 0.5 to 0.7 V, depending on how much current the DMM provides the diode.

9. Now, reverse the leads and the display should read overrange (typically shown as OL). Record the reading in Table 4-2 (use OL if it is an overrange).

10. Repeat Steps 7 to 9 for two more diodes.

11. In Fig. 4-1a, calculate the voltage across the diode and across the load resistor. Record the calculated values in Table 4-3.

12. Measure and record the value of each of the resistors used in this experiment. Build the circuit of Fig. 4-1a. Measure and record the diode and load voltages in Table 4-3.

13. Repeat Steps 11 and 12 for two more diodes.

14. In Fig. 4-1b, calculate the voltage across the diode and across the load resistor. Record the calculated values in Table 4-4.

15. Build the circuit of Fig. 4-1b. Measure and record the diode and load voltages in Table 4-4.

16. Repeat Step 15 for two more diodes.

17. Figure 4-2 shows a diode circuit. Is diode D_1 on or off? Record the answer in Table 4-5 under "D_1 normal."

18. Repeat Step 17 for the remaining diodes shown in Fig. 4-2.

19. Build the circuit shown in Fig. 4-2. Measure the current for each diode. Record the results in Table 4-5.

20. Assume that the polarity of the battery is reversed in Fig. 4-2. Determine whether each diode is on or off. Record the answers in Table 4-5 for the reverse condition.

21. Build the circuit shown in Fig. 4-2 with the polarity of the battery reversed. Measure the current for each diode. Record the results in Table 4-5.

22. In Fig. 4-3a, determine whether each diode is on or off. Record the answers in Table 4-6.

23. In Fig. 4-3b, the polarity of the voltage source is reversed. Determine whether each diode is on or off. Record the answers in Table 4-6.

24. Calculate the voltage across each diode in Fig. 4-3a. Also, calculate the load voltage. Record all values in Table 4-7.

25. Build the circuit of Fig. 4-3a. Measure and record all voltages listed in Table 4-7.

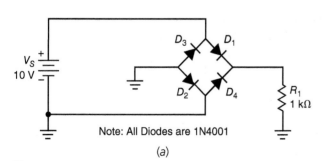

(a)

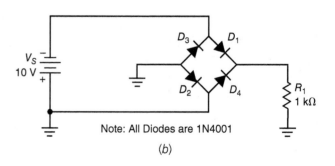

(b)

Figure 4-3

Experiment 4

Lab Partner(s) _____

PARTS USED	
Nominal Value	Measured Value
1 kΩ	
2.2 kΩ	
4.7 kΩ	
6.8 kΩ	

TABLE 4-1. OHMMETER TESTING

	Expected Resistance	Measured 1	Measured 3	Measured 3
R_F				
R_R				

TABLE 4-2. DMM TESTING

	Measured 1		Measured 2		Measured 3	
	Multisim	Actual	Multisim	Actual	Multisim	Actual
Forward						
Reverse						

TABLE 4-3. DATA FOR FORWARD BIAS

	Calculated		Measured			
			Multisim		Actual	
	V_D	V_L	V_D	V_L	V_D	V_L
Diode 1						
Diode 2						
Diode 3						

TABLE 4-4. DATA FOR REVERSE BIAS

	Calculated		Measured			
			Multisim		Actual	
	V_D	V_L	V_D	V_L	V_D	V_L
Diode 1						
Diode 2						
Diode 3						

TABLE 4-5. DIODE CONDUCTION

	Calculated				Measured			
	D_1	D_2	D_3	D_4	D_1	D_2	D_3	D_4
Normal								
Reversed								

TABLE 4-6. DIODE CONDUCTION

	D_1	D_2	D_3	D_4
Normal				
Reversed				

TABLE 4-7. DIODE AND LOAD VOLTAGES

	Calculated	Measured	
		Multisim	Actual
V_{D1}			
V_{D2}			
V_{D3}			
V_{D4}			
V_L			

Questions for Experiment 4

1. A forward-biased diode ideally appears as:
   (a) a closed switch; (b) an open switch; (c) a high resistance; (d) an ( )
   insulator.
2. A reverse-biased diode ideally appears as:
   (a) a closed switch; (b) an open switch; (c) a low resistance; (d) a ( )
   conductor.
3. When reverse-biased, a diode:
   (a) appears shorted; (b) has a low resistance; (c) has 0 V across it; ( )
   (d) appears open.
4. When a diode is tested with a DMM, the indication with reverse bias is normally:
   (a) 0.5 V; (b) 0.7 V; (c) an overrange; (d) low. ( )
5. The diode of Fig. 4-1b is:
   (a) conducting heavily; (b) reverse-biased; (c) forward-biased; (d) on. ( )
6. In Fig. 4-2, diode D_3 is:
   (a) conducting heavily; (b) reverse-biased; (c) forward-biased; (d) on. ( )
7. In Fig. 4-3a, diode D_2 is:
   (a) not conducting; (b) reverse-biased; (c) forward-biased; (d) off. ( )
8. If a diode in Fig. 4-3a has a voltage of 0.7 V across it when conducting, the load
   voltage will be: ( )
   (a) 0; (b) 8.6 V; (c) 9.3 V; (d) 10 V.
9. If a diode D_1 opens in Fig. 4-3a, the load voltage will be:
   (a) 0; (b) 8.6 V; (c) 9.3 V; (d) 10 V. ( )
10. If a diode in Fig. 4-3a has a voltage of 0.7 V across it when conducting, the voltage
    across D_4 will be: ( )
    (a) 0; (b) 8.6 V; (c) 9.3 V; (d) 10 V.

The Diode Curve

A resistor is a linear device because its voltage and current are proportional in either direction. A diode, on the other hand, is a nonlinear device because its current and voltage are not proportional. Furthermore, a diode is a unilateral device because it conducts well only in the forward direction. As a guide, a small-signal silicon diode has a dc reverse/forward resistance ratio of more than 1000:1. In this experiment, the diode currents and voltages for both forward and reverse bias will be measured. The measurement data will be used to draw the diode curve. Also included are troubleshooting, design, and application options.

GOOD TO KNOW

The cathode end of the diode is marked with a colored band.

Required Reading

Chapter 2 (Secs. 2-8 to 2-11) and Chap. 3 (Sec. 3-1) of *Electronic Principles*, 8th ed.

Equipment

1 power supply: adjustable from approximately 0 to 15 V
1 diode: 1N4148 or 1N914 (small-signal diode)
3 ½-W resistors: 220 Ω, 1 kΩ, 100 kΩ
1 DMM (digital multimeter)

Procedure

DMM DIODE TEST

1. Using the DMM as a diode tester, connect the red lead to the anode and the black lead to the cathode on the 1N4148. Measure the voltage dropped across the diode when forward biased. The diode should have approximately 0.7 V across it. Reverse the leads and measure the voltage drop across the diode. The meter should display OL.

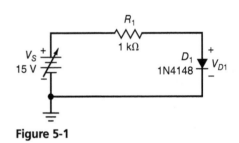

Figure 5-1

DIODE DATA

2. Measure and record the value of each of the resistors used in this experiment. Build the circuit of Fig. 5-1 using a current-limiting resistor of 1 kΩ. For each source voltage listed in Table 5-1, measure and record the diode voltage V and the diode current I.
3. Calculate and record the dc forward resistance of the diode for each current of Table 5-1.
4. Reverse the polarity of the source voltage in Fig. 5-1. For each source voltage of Table 5-2, measure and record the diode voltage V and the diode current I.
5. Calculate and record the dc reverse resistance of the diode for each source voltage of Table 5-2.

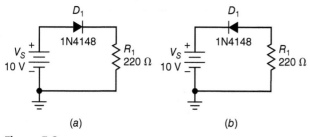

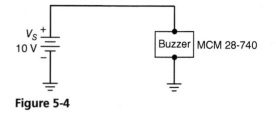

Figure 5-4

Figure 5-2

6. Use the data of Tables 5-1 and 5-2 to draw the diode curve (V versus I) on Graph 5-1.
7. The foregoing steps prove that the diode conducts easily in the forward direction and poorly in the reverse direction. Using the second approximation for a diode, calculate the diode current in Figs. 5-2a and b. Record the calculations in Table 5-3.
8. Build the circuit of Fig. 5-2a (forward bias). Measure and record the diode current in Table 5-3.
9. Build the circuit of Fig. 5-2b (reverse bias). Measure and record the diode current.

TROUBLESHOOTING

10. Build the circuit of Fig. 5-3. Using the second approximation for a diode, calculate the load voltage V_L and record in Table 5-4. Then measure and record V_L.
11. Short the diode with a jumper wire. Calculate V_L for this condition and record in Table 5-4. Measure and record V_L.
12. Remove the jumper wire. Disconnect one end of the diode. Calculate V_L and record. Next, measure and record V_L.

CRITICAL THINKING

13. Select a source voltage and a current-limiting resistance to produce 10 mA in Fig. 5-1a. (Use the resistors from earlier parts of the experiment.) Build the circuit and measure the current. Record the values of V_S and R_S, along with the measured I, in Table 5-5.

APPLICATION (OPTIONAL)

14. Connect the circuit of Fig. 5-4 using a magnetic buzzer equivalent to the MCM Electronics 28-740, which has the following specifications: 3 to 28-V dc input and 8-mA load current at 12 V dc. Notice that the buzzer is polarized, so the positive source terminal must be connected to the red lead and the negative source terminal to the black lead. An audible buzz should be present.
15. Examine Fig. 5-5. Should the buzzer work? Build the circuit of Fig. 5-5 to verify the expected result.
16. Repeat Step 15 for the circuit of Fig. 5-6.

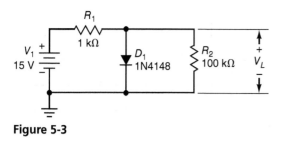

Figure 5-3

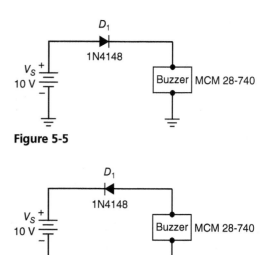

Figure 5-5

Figure 5-6

Experiment 5

Lab Partner(s) _____

PARTS USED	
Nominal Value	Measured Value
220 Ω	
1 kΩ	
100 kΩ	

TABLE 5-1. FORWARD BIAS

	Multisim		Actual		
V_S	V	I	V	I	R
0					
0.5 V					
1 V					
2 V					
4 V					
6 V					
8 V					
10 V					
15 V					

TABLE 5-2. REVERSE BIAS

	Multisim		Actual		
V_S	V	I	V	I	R
−1 V					
−5 V					
−10 V					
−15 V					

GRAPH 5-1 DIODE VOLTAGE VS. CURRENT

TABLE 5-3 DIODE CONDUCTION

	Calculated I_L	Measured I_L	
		Multisim	Actual
Fig. 5-2a			
Fig. 5-2b			

TABLE 5-4 DIODE CONDUCTION

	Calculated V_L	Measured V_L	
		Multisim	Actual
Normal Diode			
Shorted Diode			
Open Diode			

TABLE 5-5. CRITICAL THINKING

$V_S =$ _____ $R_S =$ _____ $I =$ _____

Questions for Experiment 5

1. In this experiment, the knee or offset voltage is closest to: ( )
   (a) 0.3 V;   (b) 0.7 V;   (c) 1 V;   (d) 1.2 V.
2. With forward bias, the dc resistance decreases when: ( )
   (a) current increases;   (b) diode decreases;   (c) the ratio V/I increases;
   (d) the ratio I/V decreases.
3. A diode acts like a high resistance when: ( )
   (a) its current is large;   (b) forward biased;   (c) reverse biased;
   (d) shorted.
4. Which of the following approximately describes the diode curve above the forward ( )
   knee?
   (a) it becomes horizontal;   (b) voltage increases rapidly;   (c) current increases rapidly;   (d) dc resistance increases rapidly.
5. Which of the following describes the diode curve in the reverse direction? ( )
   (a) ratio I/V is high;   (b) it becomes vertical below breakdown;   (c) dc resistance is low;   (d) current is approximately zero below breakdown.

6. Briefly describe how a diode differs from an ordinary resistor:

TROUBLESHOOTING

7. Why is the load voltage approximately 0.7 V in Fig. 5-3 when the diode is working properly?

8. Why is the load voltage slightly less than 15 V when the diode is open in Fig. 5-3?

CRITICAL THINKING

9. Why is the second approximation for the diode typically used when analyzing a low-voltage circuit and not typically used when analyzing a high-voltage circuit?

10. Optional: Instructor's question.

Diode Approximations

In the ideal or first approximation, a diode acts like a closed switch when forward biased and an open switch when reversed biased. In the second approximation, the knee voltage of the diode is included when it is forward biased. This means assuming 0.7 V across a conducting silicon diode (0.3 V for germanium). The third approximation includes the knee voltage and the bulk resistance; because of this, the voltage across a conducting diode increases as the diode current increases. In troubleshooting and design, the second approximation is usually adequate.

In this experiment, the three diode approximations will be investigated. Also included are troubleshooting, design, and application options.

GOOD TO KNOW

The widest range of values should be used to determine an accurate bulk resistance. However, both pairs of values should come from the straight part of diode's V-I curve, just right of the knee voltage.

Required Reading

Chapter 3 (Secs. 3-2 to 3-5) of *Electronic Principles*, 8th ed.

Equipment

1 power supply: adjustable from approximately 0 to 15 V
1 diode: 1N4148 or 1N914
3 ½-W resistors: two 220 Ω, 470 Ω
1 DMM (digital multimeter)

Procedure

1. Measure and record the value of each of the resistors used in this experiment. Build the circuit shown in Fig. 6-1a. Adjust the source to provide a current of 10 mA through the diode. Calculate the diode voltage V and record in Table 6-1.

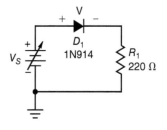

(a)

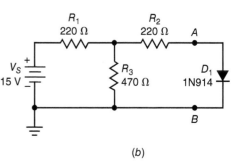

(b)

Figure 6-1

2. Measure the diode voltage V and record in Table 6-1.
3. Adjust the source to get 50 mA. Calculate the diode voltage and record in Table 6-1. Measure and record diode voltage V.
4. In this experiment, let the knee voltage be the measured diode voltage for a diode current of 10 mA. Record the knee voltage in Table 6-2. (It should be in the vicinity of 0.7 V.)
5. Calculate the bulk resistance using

$$r_B = \frac{\Delta V}{\Delta I}$$

where ΔV and ΔI are the changes in measured voltage and current in Table 6-1. Record r_B in Table 6-2.
6. Calculate the diode current in Fig. 6-1b as follows: determine the Thevenin-equivalent circuit left of the AB terminals. Then calculate the diode current with the ideal, second, and third approximations (use the V_{knee} and r_B of Table 6-2). Record the answers in Table 6-3.
7. Build the circuit shown in Fig. 6-1b. Measure and record the diode current (Table 6-3).

TROUBLESHOOTING

8. Using the second approximation for the diode, calculate the diode current in Fig. 6-1b for each of these conditions: 470 Ω shorted and open. Record the calculations in Table 6-4.
9. Measure and record the diode current in the circuit of Fig. 6-1b with the 470-Ω resistor shorted and open.

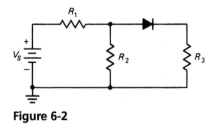

Figure 6-2

CRITICAL THINKING

10. Using the second approximation in Fig. 6-2, select values for resistors and source voltage to produce a diode current of approximately 8.9 mA. (Use the same resistance values as in Fig. 6-1b, although the resistor values can be moved.) Build the circuit using the design values and measure the diode current. Record all data listed in Table 6-5.

ADDITIONAL WORK (OPTIONAL)

11. Repeat Steps 1 to 7 for a germanium diode such as the 1N34A. Record the data on a separate piece of paper and then compare the values to those recorded for a silicon diode. How did the germanium diode differ from the silicon diode?

Experiment 6

Lab Partner(s) _____

PARTS USED Nominal Value	Measured Value
220 Ω	
220 Ω	
470 Ω	

TABLE 6-1. TWO POINTS ON THE FORWARD CURVE

I	Estimated V	Measured V	
		Multisim	Actual
10 mA			
50 mA			

TABLE 6-2. DIODE VALUES

	Multisim	Actual
$V_{knee} =$		
$r_B =$		

TABLE 6-3. DIODE CURRENT

	Measured I			
	Multisim	% Error	Multisim	Actual
Ideal I				
Second I				
Third I				

TABLE 6-4. TROUBLESHOOTING

	Estimated I	Measured I	
		Multisim	Actual
Shorted 470 Ω			
Open 470 Ω			

TABLE 6-5. CRITICAL THINKING

	Design
$V_S =$	
$R_1 =$	
$R_2 =$	
$R_3 =$	
I (diode) =	

Questions for Experiment 6

1. In this experiment, knee voltage is the diode voltage that: ( )
   (a) equals 0.3 V;   (b) equals 0.7 V;   (c) corresponds to 10 mA;   (d) corresponds to 50 mA.

2. Bulk resistance is: ( )
   (a) diode voltage divided by current;   (b) the ratio of voltage difference to current difference above the knee;   (c) the same as the dc resistance of the diode; (d) none of the foregoing.

3. The dc resistance of a silicon diode for a current of 10 mA is closest to: ( )
   (a) $2.5 \ \Omega$;   (b) $10 \ \Omega$;   (c) $70 \ \Omega$;   (d) $1 \ k\Omega$.

4. In Fig. 6-1b, the power dissipated by the diode is closest to: ( )
   (a) 0;   (b) 1.5 mW;   (c) 15 mW;   (d) 150 mW.

5. Suppose the diode of Fig. 6-1b has an $I_{F(max)}$ of 500 mA. To avoid diode damage, the source voltage can be no more than: ( )
   (a) 15 V;   (b) 50 V;   (c) 185 V;   (d) 272 V.

6. The steeper the diode curve, the smaller the bulk resistance. Explain why this is true.

TROUBLESHOOTING

7. Explain why there is no diode current when the 470-Ω resistor is shorted in Fig. 6-1b.

8. Why does the diode current increase with an open 470-Ω resistor in Fig. 6-1b?

CRITICAL THINKING

9. How many designs are possible in Step 10 of the Procedure?   ( )
   (a) 1;   (b) 2;   (c) 3;   (d) 4.

10. Optional: Instructor's question.

Input Protection

The diode is often used to limit the input voltage of an amplifier. The diode behaves as an open switch until the input voltage exceeds its barrier potential. Based on the second approximation for a silicon diode, if the input voltage is less than 0.7 V, the diode will not conduct current and therefore will appear as an open to the voltage source.

In this experiment, the use of two diodes to limit the amplitude of an input signal will be examined. This experiment will also provide an opportunity to troubleshoot and design diode input protection circuits.

GOOD TO KNOW

Some function generators produce a noisy output signal and low amplitudes. Build a voltage divider to obtain a clean small-amplitude signal if your function generator cannot produce one.

Required Reading

Chapters 2 and 3 of *Electronic Principles*, 8th ed.

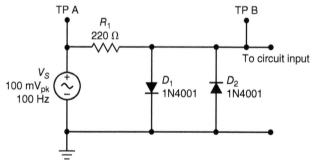

Figure SA1-1

Equipment

2   silicon diodes: 1N4001 or equivalent
1   ½-W resistor: 220Ω
1   function generator
1   oscilloscope
1   DMM (digital multimeter)

Procedure

INPUT PROTECTION

1. Sketch the circuit in Fig. SA 1-1. Using a DMM, measure and record the value of the resistor. Change the DMM mode to Diode Test, and measure and record the voltage across each of the two diodes when forward and reverse biased.

2. Based on the knowledge that a diode behaves as an open switch until the input voltage exceeds its barrier potential, the voltage measured across the diode in Fig. SA1-1 will be equal to the source voltage until the source voltage exceeds 0.7 Vp.

3. Build the circuit input protection circuit shown in Fig. SA1-1.

4. Predict and record the expected voltages at test point B for the range of input voltages listed in Table SA1-1.

5. Using the oscilloscope, while observing the input signal with channel 1 at test point A, measure the peak voltage at test point B with channel 2. Record the measured value in Table SA1-1. (*Note:* All voltage measurements are with respect to ground unless otherwise noted.)

6. Increase the amplitude of the function generator to the values shown in Table SA1-1 and record the measured peak voltages at test point B in Table SA1-1.

TROUBLESHOOTING

7. Components should not be added to or removed from the circuit while the circuit is powered. Turn off the function generator, and remove D_1 from the circuit to simulate an open D_1.

8. Turn on the function generator and adjust the amplitude of the input sine wave at test point A to 400 mVp at 100 Hz.

9. Using the oscilloscope, observe the input waveform at test point A and the output waveform at test point B.

10. Slowly increase the amplitude of the input sine wave to 5 Vp and continue to observe the voltage waveform at test point B. Record the observed results.

11. Turn off the function generator and replace D_1 and remove D_2 from the circuit. Repeat Steps 9 and 10.

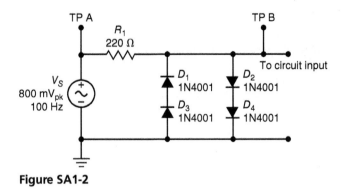

Figure SA1-2

CRITICAL THINKING

12. The voltage dropped across diodes in series is cumulative. Build the circuit shown in Fig. SA 1-2. Slowly increase the amplitude of the input sine wave from 800 mVp at test point A to the values shown in Table SA1-2. Measure the peak voltage at test point B and record the measurements in Table SA1-2

13. Based on the results observed in this experiment, design a diode input protection circuit that will limit an input signal to 3.5 Vp.

System Application 1

Lab Partner(s) _____

PARTS USED Nominal Value	Measured Value	
220 Ω		
	V_F	V_R
1N4001		
1N4001		

SCHEMATIC

TABLE SA1-1 VOLTAGE OUTPUT

Source Voltage $V_{TP\text{-}A}$ (V_p)	Expected Results $V_{TP\text{-}B}$ (V_p)	Multisim $V_{TP\text{-}B}$ (V_p)	Actual $V_{TP\text{-}B}$ (V_p)
100 mV			
300 mV			
500 mV			
700 mV			
800 mV			
900 mV			
1 V			
1.1 V			
1.2 V			
1.3 V			
2 V			
5 V			

OBSERVATIONS

TROUBLESHOOTING OBSERVATIONS

WITH D_1 REMOVED:

WITH D_2 REMOVED:

TABLE SA1-2 VOLTAGE OUTPUT

Source Voltage V_{TP-A} (V_p)	Expected Results V_{TP-B} (V_p)	Multisim V_{TP-B} (V_p)	Actual Results V_{TP-B} (V_p)
800 mV			
900 mV			
1 V			
1.1 V			
1.2 V			
1.3 V			
1.4 V			
1.5 V			
1.8 V			
2 V			
2.2 V			
5 V			

Questions for System Application 1

1. The diode behaves as a(n) _____ until the input voltage exceeds its barrier potential.
   (a) closed switch;   (b) open switch;   (c) filter;   (d) toggle switch.
2. If the input signal exceeds the breakdown voltage of the diodes, where is the remaining voltage located?
   (a) across R_1;   (b) across the load;   (c) the ac source is reduced;   (d) dissipated as heat.
3. The average barrier potential for a silicon diode is _____.
   (a) 0.7 V;   (b) 0.35 V;   (c) 50 V;   d. 0.5 V.
4. The end of the diode with a stripe around it is the _____.
   (a) anode;   (b) cathode;   (c) gate;   (d) source.
5. The barrier potential of diodes (connected cathode to anode) in series is _____.
   (a) multiplied;   (b) summed;   (c) subtracted;   (d) negated.
6. What function does the resistance value of R_1 serve?
   (a) limits V_{D1};   (b) limits V_{D2};   (c) limits I_T;   (d) limits V_{RL}.

TROUBLESHOOTING

7. In Fig. SA1-1 if D_1 opens, D_2 limits the _____ voltage to its barrier potential of approximately 700 mV.
   (a) push-over;   (b) breakdown;   (c) negative peak;   (d) positive peak.
8. In Fig. SA1-1 if D_2 opens, D_1 limits the _____ voltage to its barrier potential of approximately 700 mV.
   (a) push-over;   (b) breakdown;   (c) negative peak;   (d) positive peak.

CRITICAL THINKING

9. With an ac input signal of 400 mVp, how much current is flowing through D_1 in Fig. SA 1-2?

10. Using the second approximation for diodes, how many diodes need to be placed in series to provide input protection that will limit an input signal to 3.5 Vp?

Rectifier Circuits

The three basic rectifier circuits are the half-wave, the full-wave, and the bridge. The ripple frequency of a half-wave rectifier is equal to the input frequency, whereas the ripple frequency of a full-wave or bridge rectifier is equal to twice the input frequency. For a given transformer, the unfiltered output of the half-wave and full-wave rectifiers ideally has a dc value of slightly less than half the rms secondary voltage (45 percent), and the unfiltered output of a bridge rectifier is slightly less than the rms secondary voltage (90 percent).

In this experiment, all three types of rectifiers will be built and their input-output characteristics measured. Be especially careful in this experiment when connecting the transformer to line voltage. The transformer should have a fused line cord with all primary connections insulated to avoid electrical shock. Be sure to ask your instructor for help if you are unsure of how to properly connect the transformer.

GOOD TO KNOW

The bridge rectifier is typically sold as a single four-terminal package rather than built with four discrete diodes.

Required Reading

Chapter 4 (Secs. 4-1 to 4-4) of *Electronic Principles,* 8th ed.

Equipment

1   transformer, 12.6 V ac center-tapped (Triad F-44X or equivalent) with fused line cord
4   silicon diodes: 1N4001 (or equivalent)
1   ½-W resistor: 1 kΩ
1   DMM (digital multimeter)
1   oscilloscope

Procedure

HALF-WAVE RECTIFIER

1. Using a DMM, measure and record the value of the resistor. Change the DMM mode to Diode Test, and measure and record the voltage across each of the four diodes when forward and reverse biased. In Fig. 7-1a, the rms

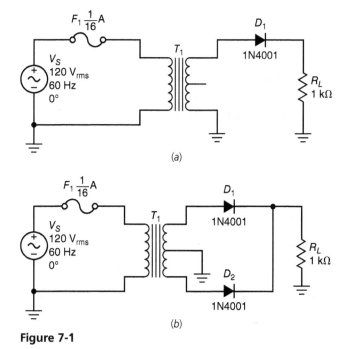

Figure 7-1

secondary output voltage is a nominal 12.6 V$_{ac}$. Using the second approximation for the diode, calculate the peak output voltage across the 1-kΩ load resistor. Also calculate the dc output voltage and ripple frequency. Record the calculations in Table 7-1.

2. Build the half-wave rectifier shown in Fig. 7-1a.
3. Measure the rms voltage across the secondary winding and record in Table 7-1.
4. Measure and record the dc load voltage.
5. Use an oscilloscope to look at the rectified voltage across the 1-kΩ load resistor. Record the peak voltage of the half-wave signal. Next, measure the period of the rectified output. Calculate the ripple frequency and record the result in Table 7-1.

FULL-WAVE RECTIFIER

6. In Fig. 7-1b, calculate and record the quantities listed in Table 7-2.
7. Build the center-tap rectifier of Fig. 7-1b.
8. Measure and record the quantities listed in Table 7-2.

BRIDGE RECTIFIER

9. In Fig. 7-2a, calculate the quantities listed in Table 7-3.
10. Build the bridge rectifier of Fig. 7-2. The bridge rectifier is typically drawn with the diodes laid out in the shape of a diamond. However, to aid in the breadboarding of the circuit, the diodes are laid out in the shape of a rectangle as shown in Fig. 7-2.
11. Measure and record the quantities listed in Table 7-3.

TROUBLESHOOTING

12. Assume that one of the diodes is open in the bridge rectifier. Calculate and record the dc output voltage and ripple frequency in Table 7-4.
13. Open one of the diodes. Measure and record the dc output voltage and ripple frequency. Restore the diode to a normal connection.
14. Assume that half of the secondary winding to the bridge rectifier is shorted (between the center tap and either end). Calculate and record the dc output voltage and ripple frequency in Table 7-4.

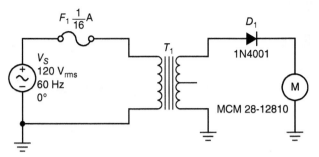

Figure 7-3

15. Simulate the foregoing short by disconnecting either end of the secondary and connecting the center tap in its place. Measure and record the dc output voltage and ripple frequency.

CRITICAL THINKING

16. Determine which modifications are needed so the bridge rectifier of Fig. 7-2 meets the following specifications: dc load voltage is approximately 5.67 V, and dc load current is approximately 20 mA. (A new load resistor will be required.)
17. Get the required load resistor and connect the modified circuit. Measure and record all the quantities listed in Table 7-5.

APPLICATION (OPTIONAL)

18. The unfiltered output of a half-wave rectifier can be used to drive a motor. The motor will respond to the dc or average value of the rectified voltage. As a demonstration, connect the circuit of Fig. 7-3 using a dc motor such as the MCM Electronics Part #: 28-12810. This dc motor has the following specifications: 6 to 24V dc input with a full-load current of 720 mA and a no-load current of 180 mA. The motor will be operated under no-load conditions since the motor is not connected to a mechanical load. After the circuit is connected, the motor's shaft can be observed rotating.
19. Reverse the diode as shown in Fig. 7-4. The motor's shaft will again turn, but this time in the opposite direction.
20. If a variac (variable line transformer) is available, it can be used to vary the line voltage to the circuit of Figs. 7-3 and 7-4. This will change the speed of the motor.

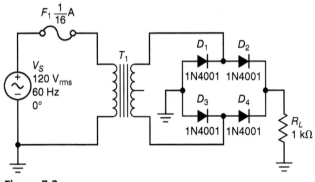

Figure 7-2

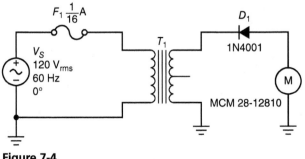

Figure 7-4

Experiment 7

Lab Partner(s) _____

PARTS USED Nominal Value	Measured Value		CALCULATIONS
1 kΩ			
	V_F	V_R	
1N4001			
1N4001			
1N4001			
1N4001			

TABLE 7-1. HALF-WAVE RECTIFIER

		Measured	
	Calculated	Multisim	Actual
RMS secondary voltage	12.6 V		
Peak output voltage			
DC output voltage			
Ripple frequency			

TABLE 7-2. FULL-WAVE RECTIFIER

		Measured	
	Calculated	Multisim	Actual
RMS secondary voltage	12.6 V		
Peak output voltage			
DC output voltage			
Ripple frequency			

TABLE 7-3. BRIDGE RECTIFIER

		Measured	
	Calculated	Multisim	Actual
RMS secondary voltage	12.6 V		
Peak output voltage			
DC output voltage			
Ripple frequency			

TABLE 7-4. TROUBLESHOOTING

| | Calculated | | Measured | | | |
| | | | Multisim | | Actual | |
	V_{dc}	f_{out}	V_{dc}	f_{out}	V_{dc}	f_{out}
Diode open						
Half-secondary short						

TABLE 7-5. CRITICAL THINKING

| | Calculated | Measured | |
		Multisim	Actual
RMS secondary voltage	6.3 V		
Peak output voltage			
DC load voltage			
DC load current			
Ripple frequency			
Load resistance			

Questions for Experiment 7

1. To measure the rms secondary voltage, it is best to use: ( )
   (a) an oscilloscope;   (b) an ammeter;   (c) a voltmeter with the common lead grounded;   (d) a floating DMM.
2. With the full-wave rectifier of this experiment, the dc load voltage was closest to: ( )
   (a) 1 V;   (b) 3 V;   (c) 6 V;   (d) 12 V.
3. The dc load voltage out of the bridge rectifier compared with the full-wave rectifier was approximately: ( )
   (a) half as large;   (b) the same;   (c) twice as large;   (d) 60 Hz.
4. Of the three rectifiers tested, the one with the largest dc output was: ( )
   (a) half-wave;   (b) full-wave;   (c) bridge;   (d) no answer.
5. The unfiltered dc output voltage from a bridge rectifier is ideally what percentage of the rms secondary voltage: ( )
   (a) 31.8;   (b) 45;   (c) 63.6;   (d) 90.
6. Explain why the bridge rectifier is the most widely used of the three types.

TROUBLESHOOTING

7. Explain why the dc output voltage and ripple frequency of a bridge rectifier drop in half when any diode opens.

8. If any diode in a bridge rectifier is shorted for any reason (solder bridge, fused diode, etc.), what will happen to the other diodes when power is applied?

CRITICAL THINKING

9. Briefly explain the design and the reasoning behind it.

10. Optional: Instructor's question.

The Capacitor-Input Filter

By connecting the output of a bridge rectifier to a capacitor-input filter, an approximately constant dc load voltage can be produced. Ideally, the filtered dc output voltage equals the peak secondary voltage. To a better approximation, the dc voltage is typically 90 to 95 percent of the peak secondary voltage with a peak-to-peak ripple of about 10 percent.

In this experiment, a bridge rectifier with a capacitor-input filter will be built. By changing load resistors and filter capacitors, the basic relations discussed in the textbook will be verified. Be especially careful in this experiment when connecting the transformer to line voltage. The transformer should have a fused line cord with all primary connections insulated to avoid electrical shock. Be sure to ask your instructor for help if you are unsure how to properly connect the transformer.

GOOD TO KNOW

Not all digital multimeters measure capacitance. Capacitance can also be measured with an LCR meter.

Required Reading

Chapter 4 (Secs. 4-6 to 4-9) of *Electronic Principles*, 8th ed.

Equipment

1 transformer: 12.6 V ac center-tapped (Triad F-44X or equivalent) with fused line cord
4 silicon diodes: 1N4001 (or equivalent)
3 ½-W resistors: 1 kΩ, 10 kΩ, 3.9 kΩ
2 capacitors: 47 μF and 470 μF (25-V rating or higher)
1 DMM (digital multimeter)
1 oscilloscope

Procedure

1. Using a DMM, measure and record the value of the resistors and the capacitors. Change the DMM mode to Diode Test, and measure and record the voltage across each of the four diodes when forward and reverse biased. Measure the resistance of the primary and secondary windings. Record in Table 8-1.

2. In Fig. 8-1, assume the rms secondary voltage is 12.6 V. Also assume $R_L = 1$ kΩ and $C = 47$ μF. Calculate and record the quantities listed in Table 8-2.
3. Build the circuit of Fig. 8-1 with $R_L = 1$ kΩ and $C = 47$ μF. The bridge rectifier is typically drawn in the shape of a diamond; however, to aid in the breadboarding of the circuit, it is shown as rectangle in Fig. 8-1.
4. Measure and record all the quantities listed in Table 8-2.
5. Repeat Steps 2 through 4 for $R_L = 1$ kΩ and $C = 470$ μF. Use Table 8-3.
6. Repeat Steps 2 through 4 for $R_L = 10$ kΩ and $C = 470$ μF. Use Table 8-4.

TROUBLESHOOTING

7. Assume that one of the diodes is open in Fig. 8-1, with $R_L = 1$ kΩ and $C = 470$ μF. Calculate the dc load voltage, ripple frequency, and peak-to-peak ripple. Record the results in Table 8-5.
8. Connect the foregoing circuit with one of the diodes open. Measure and record the quantities in Table 8-5.

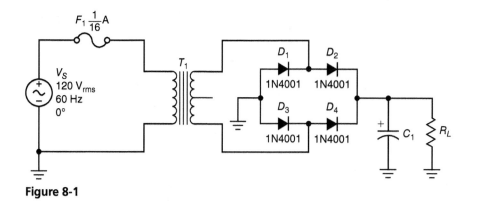

Figure 8-1

9. Assume that the filter capacitor is open in Fig. 8-1, with $R_L = 1$ kΩ and $C = 470$ μF. Calculate and record the quantities listed in Table 8-5 for this trouble.

10. Connect the circuit of Fig. 8-1 with an open filter capacitor. Measure and record the remaining quantities in Table 8-5.

CRITICAL THINKING

11. Select a filter capacitor for the circuit of Fig. 8-1 to get a peak-to-peak ripple of about 10 percent of load voltage for an R_L of 3.9 kΩ. Calculate and record the quantities of Table 8-6.

12. Build the circuit. Measure and record the quantities of Table 8-6.

GOOD TO KNOW

When dc coupled, the oscilloscope will display the ac waveform and any dc level the waveform is riding on. When ac coupled, only the ac waveform is displayed, with any dc level removed.

APPLICATION (OPTIONAL)

13. Voltage regulators are often used to provide a fixed dc voltage while minimizing the peak-to-peak ripple without the need for large-value filter capacitors. Build the circuit in Fig. 8-2. Using channels 1 and 2 of the oscilloscope, view the dc voltage and peak-to-peak ripple across C_1 and R_L.

ADDITIONAL WORK (OPTIONAL)

14. Build a half-wave rectifier with a capacitor-input filter using a filter capacitance of 470 μF and a load resistance of 1 kΩ. Measure the load voltage and peak-to-peak ripple. Record the values.

15. Repeat Step 14 for a full-wave center-tap rectifier.

16. Compare the load voltage and peak-to-peak ripple of Step 14 with Step 15. How did the results differ? What caused them to differ?

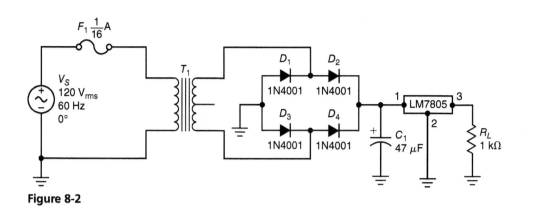

Figure 8-2

Experiment 8

Lab Partner(s) _____

PARTS USED Nominal Value	Measured Value	
1 kΩ		
3.9 kΩ		
10 kΩ		
47 μF		
470 μF		
	V_F	V_R
1N4001		
1N4001		
1N4001		
1N4001		

TABLE 8-1. TRANSFORMER RESISTANCES

	Multisim	Actual
$R_{pri} =$		
$R_{sec} =$		

TABLE 8-2. $R_L = 1$ kΩ AND $C = 47$ μF

	Calculated	Measured	
		Multisim	Actual
RMS secondary voltage	12.6 V		
Peak output voltage			
DC load voltage			
DC load current			
Ripple frequency			
Peak-to-peak ripple			

TABLE 8-3. $R_L = 1$ kΩ AND $C = 470$ μF

	Calculated	Measured	
		Multisim	Actual
RMS secondary voltage	12.6 V		
Peak output voltage			
DC load voltage			
DC load current			
Ripple frequency			
Peak-to-peak ripple			

CALCULATIONS

TABLE 8-4. $R_L = 10\ \text{k}\Omega$ AND $C = 470\ \mu\text{F}$

	Calculated	Measured	
		Multisim	Actual
RMS secondary voltage	12.6 V		
Peak output voltage			
DC load voltage			
DC load current			
Ripple frequency			
Peak-to-peak ripple			

TABLE 8-5. TROUBLESHOOTING

	Calculated			Measured					
				Multisim			Actual		
	V_{dc}	f_{out}	V_{rip}	V_{dc}	f_{out}	V_{rip}	V_{dc}	f_{out}	V_{rip}
Diode open									
Open capacitor									

TABLE 8-6. CRITICAL THINKING

	Calculated	Measured	
		Multisim	Actual
RMS secondary voltage			
Peak output voltage			
DC load voltage			
DC load current			
Ripple frequency			
Peak-to-peak ripple			

Questions for Experiment 8

1. In this experiment the dc output voltage from the capacitor-input filter was approximately equal to:　　　　　　　　　　　　　　　　　　　( )
   (a) peak primary voltage;　　(b) peak secondary voltage;　　(c) rms primary voltage;　　(d) rms secondary voltage.
2. The peak-to-peak ripple decreases when the:　　　　　　　　　　　　( )
   (a) load resistance decreases;　　(b) filter capacitor decreases;　　(c) ripple frequency decreases;　　(d) filter capacitor increases.
3. When the filter capacitor increases, the peak-to-peak ripple:　　　　　　( )
   (a) equals the secondary voltage;　　(b) remains constant;　　(c) increases;　　(d) decreases.
4. For normal operation, the ripple frequency is:　　　　　　　　　　　( )
   (a) 0;　　(b) 60 Hz;　　(c) 120 Hz;　　(d) 240 Hz.
5. When the load resistance increases, the peak-to-peak ripple:　　　　　　( )
   (a) decreases;　　(b) stays the same;　　(c) increases;　　(d) none of the foregoing.

6. Briefly explain how a capacitor-input filter works.

TROUBLESHOOTING

7. When any diode opens, the circuit of Fig. 8-1 becomes a capacitor-input filter driven by a: ( )
  (a) half-wave rectifier;   **(b)** full-wave rectifier;   **(c)** bridge rectifier;
  (d) unilateral converter.

8. Briefly explain what happens to the circuit of Fig. 8-1 when the filter capacitor opens.

CRITICAL THINKING

9. What size of capacitor was used in the design? Why was this size selected?

10. The turns ratio of the transformer is approximately 9:1. Using the transformer resistances in Table 8-1, calculate the minimum Thevenin resistance that is facing the filter capacitor. Ignore the bulk resistance of the diode and use only the primary and secondary winding resistance in the calculations.

Limiters and Peak Detectors

A positive limiter clips off positive parts of the input signal, and a negative limiter clips negative parts. In a biased limiter, the clipping level is adjustable. With a combination limiter, positive and negative parts of the signal are removed. A diode clamp is an alternative name for a limiter. Often, a diode clamp is used to protect a load from excessively high input voltages.

In this experiment, various limiters will be built and their operation observed. A peak detector, a variation of the rectifier circuits discussed earlier, will be explored. A peak detector produces a dc output voltage approximately equal to the peak voltage of the input signal.

GOOD TO KNOW

When dc coupled, the oscilloscope will display the ac waveform and any dc level the waveform is riding on. When ac coupled, only the ac waveform is displayed, with the dc level removed.

Required Reading

Chapter 4 (Sec. 4-10) of *Electronic Principles*, 8th ed.

Equipment

1   function generator
1   power supply: adjustable from approximately 0 to 15 V
2   diodes: 1N4148 or 1N914
4   ½-W resistors: 470 Ω, 1 kΩ, 10 kΩ, 100 kΩ
1   capacitor: 1 µF (10-V rating or higher)
1   DMM (digital multimeter)
1   oscilloscope

Procedure

POSITIVE LIMITER

1. In Fig. 9-1, estimate the positive and negative peak output voltages. Record in Table 9-1.
2. Using a DMM, measure and record the value of the resistors. Change the DMM mode to Diode Test.

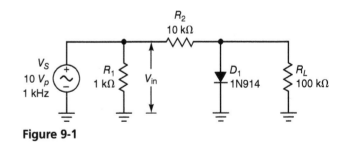

Figure 9-1

Measure and record the voltage across each of the diodes when forward and reverse biased. Build the positive limiter of Fig. 9-1. (The 1 kΩ is a dc return in case the source is capacitively coupled.) Adjust the source to get 1 kHz and 20 V peak-to-peak across the input (equivalent to a peak input of 10 V).

3. Use an oscilloscope to observe the waveform at the load resistor. A positively clipped sine wave should be displayed on the oscilloscope. Record the positive and negative peak values in Table 9-1. (The oscilloscope should be dc coupled to the load resistor.)

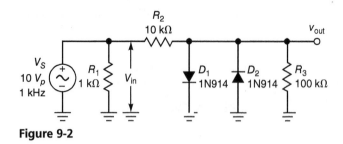

Figure 9-2

NEGATIVE LIMITER

4. In Fig. 9-1, assume the diode polarity is reversed. Record the estimates of the positive and negative output peak voltages in Table 9-1. Reverse the polarity of the diode in the circuit and observe the output waveform with an oscilloscope. It should be negatively clipped. Record the positive and negative peak values.

COMBINATION LIMITER

5. In Fig. 9-2, estimate the positive and negative peak output voltages. Record the estimates in Table 9-1. Build the combination limiter.
6. Observe the output waveform. Measure and record the positive and negative peaks.

BIASED LIMITER

7. In Fig. 9-3, estimate the output peak voltages and record in Table 9-1. Build the variable limiter of Fig. 9-3.
8. Observe the output with an oscilloscope (dc coupled). Vary the amplitude of the dc source and observe how the positive clipping level varies from a low value to a high value. If it does, write "variable" under positive peak in Table 9-1. Measure and record the negative peak.

PEAK DETECTOR

9. In Fig. 9-4, estimate the dc output voltage, ripple frequency, and peak-to-peak ripple. Use Eq. (4-10) in the textbook for the latter. Record the estimates in Table 9-2.
10. Build the peak detector of Fig. 9-4. Adjust the source to get 1 kHz and 10 V peak across the input.
11. Observe the output voltage with the oscilloscope. It should be a dc voltage with an extremely small ripple.
12. Use the DMM to measure the dc output voltage. Record this as V_{dc}.
13. Switch to the ac-coupled input on the oscilloscope and increase sensitivity until the ripple can be measured accurately. Record the ripple frequency and peak-to-peak ripple.

TROUBLESHOOTING

14. In Fig. 9-2, assume that the left diode is open. Estimate the positive and negative output peak voltages. Record the estimates in Table 9-3.
15. Build the circuit of Fig. 9-2 with the left diode open. Measure and record the output peak voltages.
16. Repeat Steps 14 and 15 for a shorted diode.

CRITICAL THINKING

17. Assume that the peak voltage is 10 V and the frequency is 5 kHz in Fig. 9-4. Select a filter capacitor (nearest standard size) that produces a peak-to-peak output ripple of approximately 0.5 V. Calculate and record all quantities listed in Table 9-4.
18. Build the circuit using the new filter capacitor. Adjust the source voltage to 20 V peak-to-peak and the frequency to 5 kHz. Measure and record all quantities listed in Table 9-4.

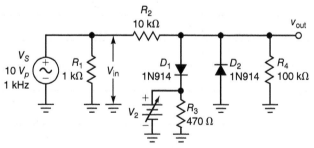

Figure 9-3

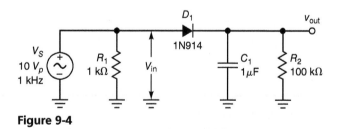

Figure 9-4

Experiment 9

Lab Partner(s) _____

PARTS USED Nominal Value	Measured Value	
470 Ω		
1 kΩ		
10 kΩ		
100 kΩ		
	V_F	V_R
1N4148 or 1N914		
1N4148 or 1N914		

OBSERVATIONS

TABLE 9-1. LIMITERS

	Estimated		Measured			
			Multisim		Actual	
	Pos peak	Neg peak	Pos peak	Neg peak	Pos peak	Neg peak
Positive limiter						
Negative limiter						
Combination limiter						
Biased limiter						

TABLE 9-2. PEAK DIRECTOR

	Estimated	Measured	
		Multisim	Actual
V_{dc}			
f_{out}			
V_{rip}			

TABLE 9-3. TROUBLESHOOTING

	Estimated		Measured			
			Multisim		Actual	
	Pos peak	Neg peak	Pos peak	Neg peak	Pos peak	Neg peak
Open diode						
Shorted diode						

TABLE 9-4. CRITICAL THINKING

	Calculated	Measured	
		Multisim	Actual
Capacitance		(no entry)	(no entry)
DC output voltage			
Ripple frequency			
Peak-to-peak ripple			

Questions for Experiment 9

1. In a negative limiter, which of these is the largest?   ( )
   (a) positive peak;   (b) negative peak;   (c) knee voltage;   (d) crossover voltage.
2. The combination limiter of Fig. 9-2:   ( )
   (a) puts out a small sine wave;   (b) generates a small squarish wave;   (c) has an adjustable clipping level;   (d) has an output proportional to the input.
3. When the dc source of Fig. 9-3 varies from 0 to 15 V, the positive output peak varies   ( )
   from roughly:
   (a) 0 to $V_P/2$;   (b) 0 to V_P;   (c) 0 to $2V_P$;   (d) 0 to 0.7 V.
4. In the combination limiter of Fig. 9-2, which diode approximation is the most reasonable   ( )
   compromise?
   (a) ideal;   (b) second;   (c) third;   (d) fourth.
5. The peak-to-peak ripple out of the peak detector of Fig. 9-4 was approximately what   ( )
   percent of the dc output voltage?   ( )
   (a) 1%;   (b) 5%;   (c) 10%;   (d) 20%.
6. Briefly explain the operation of the biased combination clipper (Fig. 9-3).

TROUBLESHOOTING

7. Explain why each trouble in Table 9-3 produces the recorded outputs.

8. When troubleshooting a peak detector like Fig. 9-4, if the output is a half-wave rectified sine wave, what is the trouble?

CRITICAL THINKING

9. Which diode approximation appears to be the best compromise for designing peak detectors? Explain why it is the best solution.

10. What advantage would a germanium high-frequency signal diode have over a similar silicon diode used as a peak detector? Why is this an advantage?

DC Clampers and Peak-to-Peak Detectors

In a dc clamper, a capacitor is charged to approximately the peak input voltage V_P. Depending on the polarity of the charge, the output voltage has a dc component equal to the positive or negative peak input voltage. The output of a positive clamper ideally swings from 0 to $+2V_P$, whereas the output of a negative clamper swings from 0 to $-2V_P$.

A peak-to-peak detector is a cascaded connection of a dc clamper and a peak detector. The dc clamper ideally produces an output that swings from 0 to $2V_P$, and the peak detector produces a dc output of approximately $2V_P$. Since the final dc output equals the peak-to-peak input voltage, the overall circuit is called a peak-to-peak detector.

If a signal source is capacitively coupled, the problem of the dc return may arise with diode and transistor circuits. When the source has to supply more current on one half-cycle than the other, its coupling capacitor will charge to approximately the peak of the source voltage. Because of this, an unwanted dc clamping of the source signal will occur. To eliminate this unwanted clamping, a dc return is required. It discharges the coupling capacitor and prevents a dc shift of the output signal.

GOOD TO KNOW

To observe a small ac signal riding on a dc level, set the oscilloscope to ac coupling. This will remove the dc level from the display and allow a smaller volts per division setting to be used.

Required Reading

Chapter 4 (Sec. 4-11) of *Electronic Principles*, 8th ed.

Equipment

1   function generator
2   diodes: 1N4148 or 1N914
4   ½-W resistors: 1 kΩ, 10 kΩ, 47 kΩ, 100 kΩ
2   capacitors: 1 μF (20-V rating or higher)
1   DMM (digital multimeter)
1   oscilloscope

Procedure

POSITIVE CLAMPER

1. In Fig. 10-1, estimate the positive and negative peaks of the output voltage. Record in Table 10-1.
2. Using a DMM, measure and record the values of the resistors and capacitors. Change the DMM mode to Diode Test. Measure and record the voltage across each of the diodes when forward and reverse biased. Build the positive clamper of Fig. 10-1. Adjust the source to provide a 1-kHz and, 10-V_p input voltage.

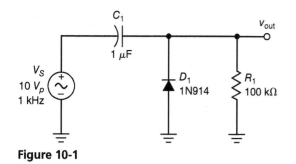

Figure 10-1

3. With the oscilloscope set for dc coupling, observe the output. It should be a positively clamped sine wave. Measure and record the positive and negative peaks in Table 10-1.

4. Leave the oscilloscope on the output and vary the input voltage. Observe how the negative peak is clamped near zero while the positive peak moves up and down.

NEGATIVE CLAMPER

5. Assume the polarity of the diode in Fig. 10-1 is reversed. Estimate and record the output peaks in Table 10-1.

6. Reverse the polarity of the diode in the constructed circuit. Measure and record the output peaks.

PEAK-TO-PEAK DETECTOR

7. Estimate the dc output voltage and peak-to-peak ripple in Fig. 10-2. Use Eq. (4-10) for the latter. Record in Table 10-2.

8. Build the peak-to-peak detector of Fig. 10-2. Adjust the source to get 1 kHz and 20 V peak-to-peak across the input.

9. Observe the voltage across the first diode. It should be a positively clamped signal.

10. Observe the output. It should be a dc voltage with a small ripple. Measure the dc output voltage with a DMM and record in Table 10-2.

11. Switch the oscilloscope to ac coupled and a smaller volts per division setting to measure the ripple. Record V_{rip}.

DC RETURN

12. In Fig. 10-3, the inside of the dashed box simulates a capacitively coupled source. The 1-kΩ resistor is a dc return. Estimate and record the positive peak output voltage (Table 10-3). Visualize the dc return open; estimate and record the positive-peak output voltage.

13. Build the circuit of Fig. 10-3. Adjust the source to get 1 kHz and 20 V peak-to-peak across the 1-kΩ resistor.

14. Observe the output with the oscilloscope. It should be a half-wave signal. Measure and record the peak value in Table 10-3.

15. Disconnect the 1 kΩ. Measure and record the output peak value.

TROUBLESHOOTING

16. In Fig. 10-2, assume capacitor C_1 is open. Estimate and record the dc output voltage in Table 10-4.

17. Build the circuit with the foregoing trouble. Measure and record the dc output voltage.

18. Repeat Steps 16 and 17 for each of the remaining troubles listed in Table 10-2.

CRITICAL THINKING

19. The frequency is changed to 2.5 kHz and the load resistor to 47 kΩ in Fig. 10-2. Select a value of output filter capacitance (nearest standard size) that produces a peak-to-peak ripple of approximately 0.1 V. Calculate all quantities listed in Table 10-5.

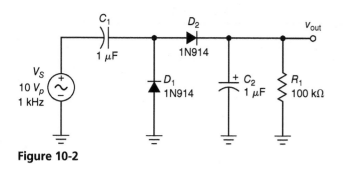

Figure 10-2

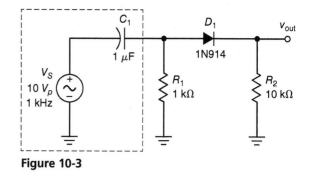

Figure 10-3

Experiment 10

Lab Partner(s) _____

PARTS USED Nominal Value	Measured Value	
1 kΩ		
10 kΩ		
47 kΩ		
100 kΩ		
	V_F	V_R
1N4148 or 1N914		
1N4148 or 1N914		

OBSERVATIONS

TABLE 10-1. CLAMPERS

	Estimated		Measured			
			Multisim		Actual	
	Pos peak	Neg peak	Pos peak	Neg peak	Pos peak	Neg peak
Positive clamper						
Negative clamper						

TABLE 10-2. PEAK-TO-PEAK DETECTOR

	Estimated	Measured	
		Multisim	Actual
V_{dc}			
V_{rip}			

TABLE 10-3. DC RETURN

	Estimated V_p	Measured V_p	
		Multisim	Actual
With dc return			
Without dc return			

TABLE 10-4. TROUBLESHOOTING

	Estimated V_{dc}	Measured V_{dc}	
		Multisim	Actual
Open C_1			
Shorted C_1			
Open C_2			
Shorted C_2			
Open D_1			
Shorted D_1			
Open D_2			
Shorted D_2			

TABLE 10-5. CRITICAL THINKING

	Calculated	Measured			
		Multisim	% Error	Actual	% Error
Capacitance		(no entry)	(no entry)	(no entry)	(no entry)
DC output voltage					
Ripple frequency					
Peak-to-peak ripple					

Questions for Experiment 10

1. If the diode of Fig. 10-1 is reversed, the output will be: ( )
   (a) positively clamped;   (b) negatively clamped;   (c) half-wave rectified;
   (d) peak rectified.
2. If V_P is 10 V in Fig. 10-2, the maximum positive voltage across the first diode is   ( )
   approximately:
   (a) 5 V;   (b) 10 V;   (c) 15 V;   (d) 20 V.
3. If V_P is 10 V in Fig. 10-2, the dc output voltage is ideally: ( )
   (a) 0 V;   (b) 5 V;   (c) 10 V;   (d) 20 V.
4. In Fig. 10-2, the peak-to-peak ripple is approximately what percentage of dc output voltage?   ( )
   put voltage?
   (a) 0;   (b) 1%;   (c) 5%;   (d) 63.6%.
5. When the dc return of Fig. 10-3 is disconnected, which of the following is true?   ( )
   (a) the capacitor charges to approximately $2V_P$;   (b) current flows easily in
   the reverse diode direction;   (c) the diode conducts briefly near each positive
   peak;   (d) the diode eventually stops conducting.
6. Explain how a positive dc clamper works.

TROUBLESHOOTING

7. What dc output voltage was measured when C_2 was opened in Fig. 10-2? Explain why this happened.

8. A group of technicians are gathered around a circuit that works with one signal generator but not with another. No one can figure out why one generator works but not the other. Explain what is probably happening.

9. Optional: Instructor's question.

10. Optional: Instructor's question.

Experiment 11

Voltage Doublers

A voltage multiplier produces a dc voltage equal to a multiple of the peak input voltage. Voltage multipliers are useful with high-voltage/low-current loads. With a voltage doubler, the dc output voltage is twice that of a standard peak rectifier. This is useful when producing high voltages (several hundred volts or more) because higher secondary voltages result in bulkier transformers. At some point, a designer may prefer to use voltage doublers instead of bigger transformers. With a voltage tripler, the dc voltage is approximately three times the peak input voltage. As the multiple increases, the peak-to-peak ripple gets worse.

In this experiment, half-wave and full-wave voltage doublers will be built. The dc output voltage and peak-to-peak ripple of these circuits will be measured to verify the operation described in the textbook. The transformer should have a fused line cord with all primary connections insulated to avoid electrical shock. Be sure to ask your instructor for help, if you are unsure of how to properly connect the transformer.

GOOD TO KNOW

Oscilloscope probes are available in 1:1, 10:1, 100:1, and 1000:1 configurations. The 100:1 and 1000:1 probes are used for high voltages.

Required Reading

Chapter 4 (Sec. 4-12) of *Electronic Principles*, 8th ed.

Equipment

1   transformer: 12.6 V ac center-tapped (Triad F-44X or equivalent) with fused line cord
2   silicon diodes: 1N4001 (or equivalent)
2   ½-W resistors: 1 kΩ, 3.9 kΩ
2   capacitors: 470 μF (25-V rating or higher)
1   DMM (digital multimeter)
1   oscilloscope

Procedure

HALF-WAVE DOUBLER

1. Using a DMM, measure and record the values of the resistors and capacitors. Change the DMM mode to Diode Test. Measure and record the voltage across each of the diodes when forward and reverse biased. Measure the resistance of the primary and secondary windings of the transformer. Record in Table 11-1.
2. In Fig. 11-1, assume that the rms secondary voltage is 12.6 V. Calculate and record the quantities listed in Table 11-2. Use Eq. (4-10) in the textbook to calculate the peak-to-peak ripple.

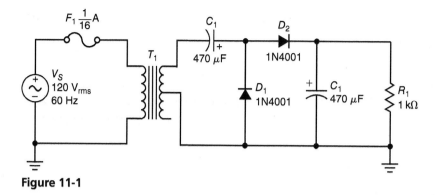

Figure 11-1

3. Build the circuit.
4. Measure and record all the quantities listed in Table 11-2.

FULL-WAVE DOUBLER

5. Repeat Steps 2 through 4 for the full-wave doubler of Fig. 11-2. Use Table 11-3 to record the data. When calculating the peak-to-peak ripple, notice that the load resistor is in parallel with two capacitors in series.

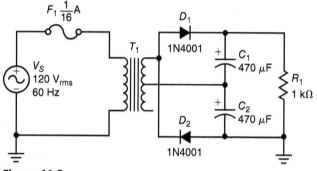

Figure 11-2

TROUBLESHOOTING

6. Assume that capacitor C_1 is open in Fig. 11-1.
7. Estimate the dc load voltage, ripple frequency, and peak-to-peak ripple. Record the estimated values in Table 11-4.
8. Build the circuit with the foregoing trouble. Measure and record the quantities of Table 11-4.
9. Assume that diode D_2 is open in Fig. 11-1. Repeat Steps 7 and 8.

CRITICAL THINKING

10. Select a filter capacitor (nearest standard size) for the circuit of Fig. 11-1 to get a peak-to-peak ripple of approximately 10 percent of load voltage for an R_L of 3.9 kΩ. Calculate and record the quantities of Table 11-5. Record the design value for capacitance here:

$$C =$$

11. Connect the circuit. Measure and record the quantities of Table 11-5.

Experiment 11

Lab Partner(s) _____

PARTS USED Nominal Value	Measured Value	
1 kΩ		
3.9 kΩ		
470 μF		
470 μF		
	V_F	V_R
1N4148 or 1N914		
1N4148 or 1N914		

TABLE 11-1. TRANSFORMER RESISTANCES

R_{pri}	
R_{sec}	

TABLE 11-2. HALF-WAVE DOUBLER

	Calculated	Measured	
		Multisim	Actual
Half rms secondary voltage			
DC output voltage			
Ripple frequency			
Peak-to-peak ripple			

TABLE 11-3. FULL-WAVE DOUBLER

	Calculated	Measured	
		Multisim	Actual
Half rms secondary voltage			
DC output voltage			
Ripple frequency			
Peak-to-peak ripple			

TABLE 11-4. TROUBLESHOOTING

	Estimated			Measured					
				Multisim			Actual		
	V_{dc}	f_{out}	V_{rip}	V_{dc}	f_{out}	V_{rip}	V_{dc}	f_{out}	V_{rip}
Open C_1									
Open D_2									
Open C_2									

TABLE 11-5. CRITICAL THINKING

	Calculated	Measured I			
		Multisim	% Error	Actual	% Error
Half rms secondary voltage					
DC output voltage					
Ripple frequency					
Peak-to-peak ripple					

Questions for Experiment 11

1. In this experiment, the dc output voltage from the half-wave doubler was approxi-   ( )
   mately equal to:
   (a) peak primary voltage;   **(b)** rms secondary voltage;   **(c)** double the peak
   secondary voltage;   **(d)** double the peak voltage driving the half-wave doubler.
2. The ripple frequency of a half-wave doubler was: ( )
   (a) 60 Hz;   **(b)** 120 Hz;   **(c)** 240 Hz;   **(d)** 480 Hz.
3. The full-wave doubler has a ripple frequency of: ( )
   (a) 60 Hz;   **(b)** 120 Hz;   **(c)** 240 Hz;   **(d)** 480 Hz.
4. The peak-to-peak ripple of a full-wave doubler compared with a half-wave dou-   ( )
   bler is:
   (a) half;   **(b)** the same;   **(c)** twice as much.
5. In Fig. 11-1, D_1 and C_1 act like a: ( )
   (a) positive clamper;   **(b)** negative clamper;   **(c)** peak detector;
   (d) diode clamp.
6. Briefly explain how the full-wave doubler of Fig. 11-2 works.

TROUBLESHOOTING

7. Explain why the peak-to-peak ripple is so large for an open C_2 in Table 11-4.

8. Suppose either filter capacitor in Fig. 11-2 is shorted. Explain what happens to the nearest
   diode.

CRITICAL THINKING

9. Justify the design; that is, why was the filter capacitor selected?

10. Assume that the primary resistance is 30 Ω and the secondary resistance is 1 Ω in Fig. 11-2.
    The primary voltage is 115 V, and the secondary voltage is 12.6 V. What is the Thevenin resis-
    tance facing either filter capacitor? Ignore the bulk resistance of the diodes in this calculation.

The Zener Diode

Ideally, a zener diode is equivalent to a dc source when operating in the breakdown region. It is important to remember that the zener diode must be reverse biased to operate in the breakdown region. To a second approximation, it is like a dc source with a small internal impedance. Its main advantage is the approximately constant voltage appearing across it. In this experiment, the operation of the zener diode will be explored. The zener diode voltage will be measured over a range of input voltages, and the data will be used to calculate the zener resistance.

GOOD TO KNOW

The cathode end of the zener diode is marked with a colored band.

Anode ◢◣ Cathode

Required Reading

Chapter 5 (Secs. 5-1 to 5-7) of *Electronic Principles*, 8th ed.

Equipment

1 power supply: adjustable from approximately 0 to 15 V
2 zener diodes: 1N5234B
1 ½-W resistor: 180 Ω
1 DMM (digital multimeter)

Procedure

ZENER VOLTAGE

1. Using a DMM, measure and record the value of the resistor. Change the DMM mode to Diode Test. Measure and record the voltage across the zener diode when forward and reverse biased. The diode should test similar to a rectifier diode.
2. The 1N5234B has a nominal zener voltage of 6.2 V. In Fig. 12-1, estimate and record the output voltage for each input voltage listed in Table 12-1.
3. Build the circuit of Fig. 12-1. Measure and record the output voltage for each input voltage of Table 12-1.

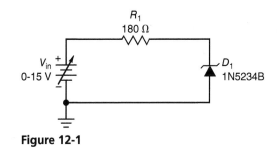

Figure 12-1

ZENER RESISTANCE

4. With the data of Table 12-1, calculate and record the zener current in Fig. 12-1 for each entry of Table 12-2.
5. Using the equation $R_z = \dfrac{\Delta V_z}{\Delta I_z}$, calculate the zener resistance for $V_{in} = 10$ V. (Use the voltage and current changes between 8 and 12 V.)
6. Calculate and record the zener resistance for $V_{in} = 12$ V.

CURVE TRACER

7. If a curve tracer is available, display the forward and reverse zener curves.

TROUBLESHOOTING

8. In Fig. 12-1, assume that V_{in} is 9 V and estimate the output voltage for a shorted zener diode. Record the answer in Table 12-3.
9. Estimate and record the output voltage for an open zener diode.
10. Estimate and record V_{out} for an open resistor.
11. Assume the polarity of the zener diode is reversed. Estimate and record the output voltage for this trouble.
12. Connect the circuit with each of the foregoing troubles. Measure and record V_{out} for a V_{in} of 9 V.

CRITICAL THINKING

13. Select a current-limiting resistor to produce a zener current of approximately 16.5 mA when V_{in} is 14 V. Record the resistor value at the top of Table 12-4. Build the circuit with the design value of R_S. Measure and record the output voltage for each input voltage listed in Table 12-4.

14. Calculate and record the zener current for each input voltage of Table 12-4. Calculate and record the zener resistance for an input voltage of 12 V.

APPLICATION (OPTIONAL)

15. When two zener diodes are placed in series as shown in Fig. 12-2, the total voltage dropped across both diodes is the sum of their rated zener voltages. Build the circuit in Fig. 12-2 and compare the measured V_{out} to the estimated value.

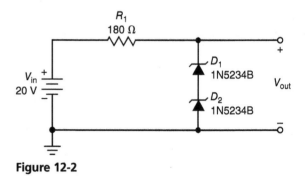

Figure 12-2

Experiment 12

Lab Partner(s) _____

PARTS USED Nominal Value	Measured Value	
180 Ω		
	V_F	V_R
1N5234B		
1N5234B		

TABLE 12-1. DATA FOR ZENER DIODE

V_{in}	Estimated V_{out}	Measured V_{out}	
		Multisim	Actual
0			
2 V			
4 V			
6 V			
8 V			
10 V			
12 V			
14 V			

TABLE 12-2. ZENER RESISTANCE

V_{in}	Calculated I_Z	Calculated R_Z
0		(no entry)
2 V		(no entry)
4 V		(no entry)
6 V		(no entry)
8 V		(no entry)
10 V		
12 V		
14 V		(no entry)

TABLE 12-3. TROUBLESHOOTING

Trouble	Estimated V_{dc}	Measured V_{dc}	
		Multisim	Actual
Shorted diode			
Open diode			
Open resistor			
Reversed diode			

TABLE 12-4. CRITICAL THINKING: $R_S =$ _____

V_{in}	Measured V_{out}		Calculated I_Z	Calculated R_Z
	Multisim	Actual		
10 V				(no entry)
12 V				
14 V				(no entry)

Questions for Experiment 12

1. In Fig. 12-1, the zener current and the current through the 180-Ω resistor are: ( )
   (a) equal;   (b) almost equal;   (c) much different.
2. The zener diode starts to break down when the input voltage is approximately: ( )
   (a) 4 V;   (b) 6 V;   (c) 8 V;   (d) 10 V.
3. When V_{in} is less than 6 V, the output voltage is: ( )
   (a) approximately constant;   (b) negative;   (c) the same as the input.
4. When V_{in} is greater than 8 V, the output voltage is: ( )
   (a) approximately constant;   (b) negative;   (c) the same as the input.
5. The calculated zener resistances were closest to: ( )
   (a) 1 Ω;   (b) 2 Ω;   (c) 7 Ω;   (d) 20 Ω.
6. Explain why the zener diode is called a constant-voltage device.

TROUBLESHOOTING

7. Explain the measured value of output voltage when the zener diode was open.

8. Explain the measured value of output voltage when the zener diode was reversed.

CRITICAL THINKING

9. Describe the criteria for the selection of the current-limiting resistance.

10. Optional. Instructor's question.

The Zener Regulator

In a zener voltage regulator, a load resistor is in parallel with a zener diode. As long as the zener diode operates in the breakdown region, the load voltage is approximately constant and equal to the zener voltage. In a stiff zener regulator, the zener resistance is less than 1/100 of the series resistance and less than 1/100 of the load resistance. By meeting the first condition, a zener regulator attenuates the input ripple by a factor of at least 100. By meeting the second condition, a zener regulator appears like a stiff voltage source to the load resistance.

In this experiment, a split supply with regulated positive and negative output voltages will be built. The operation of a zener diode will be verified by measuring the output voltages of the power supply. The transformer should have a fused line cord with all primary connections insulated to avoid electrical shock. Be sure to ask your instructor for help, if you are unsure of how to properly connect the transformer.

GOOD TO KNOW

The cathode end of the zener diode is marked with a colored band.

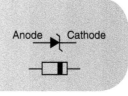

Required Reading

Chapter 5 (Secs. 5-1 to 5-7) of *Electronic Principles*, 8th ed.

Equipment

1   center-tapped transformer, 12.6 V ac (Triad F-44X or equivalent) with fused line cord
4   silicon diodes: 1N4001 (or equivalent)
2   zener diodes: 1N5234B
8   ½-W resistors: two 150 Ω, two 470 Ω, two 4.7 kΩ, two 47 kΩ
2   capacitors: 470 μF (25-V rating or better)
1   DMM (digital multimeter)
1   oscilloscope

Procedure

SPLIT SUPPLY

1. A 1N5234B has a nominal zener voltage of 6.2 V. In Fig. 13-1, calculate the input and output voltages for each zener regulator. (The input voltages are across the filter capacitors.) Record the answers in Table 13-1.

2. Using a DMM, measure and record the value of the resistors and capacitors. Change the DMM mode to Diode Test, and measure and record the voltage across each of the four diodes when forward and reverse biased. Build the split supply of Fig. 13-1 without the 470-Ω load resistor.

3. Measure the input and output voltages of each zener regulator. Record the data in Table 13-1.

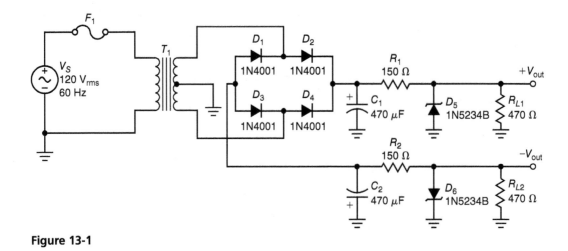

Figure 13-1

VOLTAGE REGULATION

4. Estimate and record the output voltages in Fig. 13-1 for each of the load resistors listed in Table 13-2.
5. Add the 470-Ω load resistor to the circuit, completing the circuit shown in Fig. 13-1. Measure and record the output voltages for each of the load resistances listed in Table 13-2.

RIPPLE ATTENUATION

6. For each load resistance listed in Table 13-3, calculate and record the peak-to-peak ripple across the upper filter capacitor of Fig. 13-1. Also calculate and record the peak-to-peak ripple at the positive output. (Assume a zener resistance of 7 Ω.)
7. For each load resistance of Table 13-3, measure and record the peak-to-peak ripple at the input and output of the positive zener regulator.

TROUBLESHOOTING

8. Assume that the center tap of Fig. 13-1 is open.
9. Estimate the output voltages for the foregoing trouble. Record the answers in Table 13-4.
10. Connect the circuit with the foregoing trouble. Measure and record the output voltages. Remove the trouble.
11. Repeat Steps 9 and 10 for the other troubles listed in Table 13-4.

CRITICAL THINKING

12. Design a two-stage voltage regulator similar to Fig. 5-7 of the textbook to meet these specifications: preregulator output is a nominal +12.4 V, final output is a nominal +6.2 V, current in preregulator series resistor is 40 mA, current in final series resistor is 20 mA, and ripple attenuation is at least 300. Assume a zener resistance of 7 Ω for each diode. Use the 1N5234B zener diodes and any additional resistors as required. Draw the final design underneath Table 13-5.
13. Calculate and record the dc voltage and peak-to-peak ripple at the preregulator input, regulator input, and final output (Table 13-5).
14. After checking with the instructor about the safety of the design, build the circuit with a load resistance of 470 Ω. Measure all dc voltages and ripples listed in Table 13-5. Record the data.

ADDITIONAL WORK (OPTIONAL)

15. Have another student insert one of the following troubles into the circuit: *open* any diode, resistor, capacitor, fuse, or connecting wire. Use only voltage readings of a DMM or oscilloscope to troubleshoot.
16. Repeat Step 15 several times until open components can be quickly located.

Experiment 13

Lab Partner(s) _____

PARTS USED Nominal Value	Measured Value	PARTS USED	Measured Value	
			V_F	V_R
150 Ω				
150 Ω		1N4001		
470 Ω		1N4001		
470 Ω		1N4001		
4.7 kΩ		1N4001		
4.7 kΩ		1N5234B		
47 kΩ		1N5234B		
47 kΩ				
470 μF				
470 μF				

TABLE 13-1. SPLIT SUPPLY

	Calculated		Measured			
			Multisim		Actual	
	V_{in}	V_{out}	V_{in}	V_{out}	V_{in}	V_{out}
Positive regulator						
Negative regulator						

TABLE 13-2. VOLTAGE REGULATION

	Estimated		Measured			
			Multisim		Actual	
R_L	$+V_{out}$	$-V_{out}$	$+V_{out}$	$-V_{out}$	$+V_{out}$	$-V_{out}$
470 Ω						
4.7 kΩ						
47 kΩ						

TABLE 13-3. RIPPLE

	Calculated V_{rip}		Measured V_{rip}			
			Multisim		Actual	
R_L	In	Out	In	Out	In	Out
470 Ω						
47 kΩ						

TABLE 13-4. TROUBLESHOOTING

| | Estimated | | Measured | | | |
| | | | Multisim | | Actual | |
R_L	$+V_{out}$	$-V_{out}$	$+V_{out}$	$-V_{out}$	$+V_{out}$	$-V_{out}$
Open CT						
Open D_1						
Open D_6						

TABLE 13-5. CRITICAL THINKING

| | Calculated | | Measured | | | | | |
| | | | Multisim | | | Actual | | |
	V_{dc}	V_{rip}	V_{dc}	% Error	V_{rip}	V_{dc}	% Error	V_{rip}
Preregulator input								
Preregulator input								
Regulator output								

Draw the design here:

Questions for Experiment 13

1. A split supply has: ( )
   (a) only one output voltage;   **(b)** only a positive output voltage;   **(c)** only a negative output voltage;   **(d)** positive and negative outputs.
2. The value of V_{in} to the positive zener regulator is closest to: ( )
   (a) 5 V;   **(b)** 10 V;   **(c)** 15 V;   **(d)** 20 V.
3. When R_L increases in Table 13-2, the measured positive output voltage: ( )
   (a) decreases slightly;   **(b)** remains the same;   **(c)** increases slightly.
4. Theoretically, the positive zener regulator of Fig. 13-1 attenuates the ripple by a   ( )
   factor of approximately:
   (a) 10;   **(b)** 20;   **(c)** 50;   **(d)** 100.
5. The current through either series resistor of Fig. 13-1 is closest to: ( )
   (a) 5 mA;   **(b)** 10 mA;   **(c)** 15 mA;   **(d)** 20 mA.
6. Explain how the positive zener regulator of Fig. 13-1 works.   ( )

TROUBLESHOOTING

7. Explain why the circuit of Fig. 13-1 continued to work even though the center tap was open.

8. Explain why the circuit of Fig. 13-1 still works with an open D_2.

CRITICAL THINKING

9. Explain why the measured ripples did not agree exactly with the calculated ripples in the design.

10. Optional. Instructor's question.

Voltage Regulation

The typical dc power supply is composed of four sections: transformer, rectifier, filter, and regulator. The most common form of rectification is the bridge rectifier. A large-value filter capacitor is commonly used to smooth out the rectified voltage. A zener diode or a three-terminal voltage regulator is found in the Regulator section of Fig. SA2-1.

AC Input

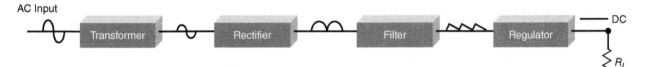

Figure SA2-1

In this experiment, the use of a zener diode and a three-terminal voltage regulator utilized to regulate the output voltage in a power supply will be examined. This experiment will also provide an opportunity to troubleshoot and design a dc power supply circuit. The transformer should have a fused line cord with all primary connections insulated to avoid electrical shock. Be sure to ask your instructor for help, if you are unsure of how to properly connect the transformer.

GOOD TO KNOW

When dc coupled, the oscilloscope will display the ac waveform and any dc level the waveform is riding on. When ac coupled, only the ac waveform is displayed, with any dc level removed.

Required Reading

Chapters 4 and 5 of *Electronic Principles*, 8th ed.

Equipment

1   transformer: 12.6 V ac center-tapped (Triad F-44X or equivalent) with fused line cord
4   silicon diodes: 1N4001 (or equivalent)

1   zener diode: 1N5231B (or equivalent)
1   voltage regulator: LM7805C (or equivalent)
3   ½-W resistors: 220 Ω, 1 kΩ, 10 kΩ
6   capacitors: 0.1 μF, 0.33 μF, 1 μF, 4.7 μF, 10 μF, 33 μF (25-V rating or higher)
1   DMM (digital multimeter)
1   oscilloscope

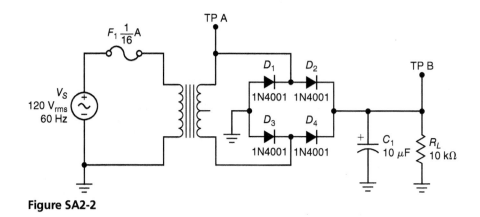

Figure SA2-2

Procedure

VOLTAGE REGULATION

1. Sketch the pinout of the LM7805C voltage regulator. Using a DMM, measure and record the value of the resistors and capacitors. Change the DMM mode to Diode Test, and measure and record the voltage across each of the four silicon diodes when forward and reverse biased.

2. Build the dc power supply circuit, as shown in Fig. SA2-2.

3. Using the formula $V_R = V_{P(\text{out})} (1 - \varepsilon^{-t/R_L C})$ where $t = 8.33$ ms, calculate the expected ac peak-to-peak ripple voltage across the load resistor R_L. Record the calculated value on Table SA2-1.

4. Using the oscilloscope, while observing the input signal at test point A with channel 1, measure the dc voltage and ac peak-to-peak ripple voltage across the load resistor R_L at test point B with channel 2. Record the measured values in Table SA2-1. (*Note:* Use dc coupling for the dc voltage measurement and ac coupling for the ac ripple voltage measurement.)

5. Be sure to always turn off the power to the circuit before adding or removing components. Add the series resistor R_1 and the zener diode to the circuit, as shown in Fig. SA2-3. Reapply power to the circuit.

6. Using the oscilloscope, while observing the input signal at test point A with channel 1, measure the dc voltage and ac peak-to-peak ripple voltage at both test points B and C with channel 2. Record the measured values in Table SA2-2. If a four-channel oscilloscope is used, view all three test points at the same time. (*Note:* Use dc coupling for the dc voltage measurement and ac coupling for the ac ripple voltage measurement.)

7. Be sure to always turn off the power to the circuit before adding or removing components. Remove the 220 Ω resistor and the zener diode and replace them with the voltage regulator, as shown in Fig. SA 2-4. Reapply power to the circuit.

8. Using the oscilloscope, while observing the input signal at test point A with channel 1, measure both the dc voltage and ac peak-to-peak ripple voltage at test points B and C with channel 2. Record the measured values in Table SA2-3. If a four-channel oscilloscope is used, view all three test points at the same time. (*Note:* Use dc coupling for the dc voltage measurement and ac coupling for the ac ripple voltage measurement.)

TROUBLESHOOTING

9. As electrolytic capacitors age, their actual capacitance value tends to decrease. Simulate a failing capacitor in Fig. SA2-3 by replacing C_1 with a 4.7 μF capacitor. Using the oscilloscope, while observing the input signal at test point A with channel 1, measure both the dc voltage and ac peak-to-peak ripple voltage at test points B and C with channel 2. Record the measured values in Table SA2-4.

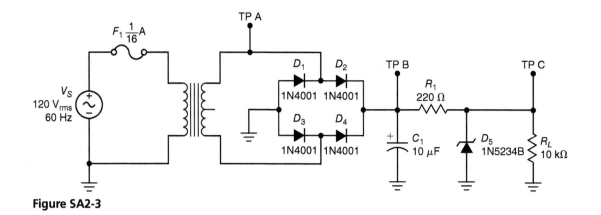

Figure SA2-3

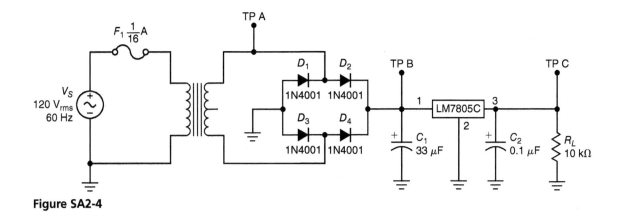

Figure SA2-4

10. Further simulate a failing capacitor in Fig. SA2-3 by replacing C_1 with a 1 μF capacitor. Using the oscilloscope, while observing the input signal at test point A with channel 1, measure both the dc voltage and ac peak-to-peak ripple voltage at test points B and C with channel 2. Record the measured values in Table SA2-4.

11. Record the observed ac ripple voltage as the capacitor C_1 is reduced in value.

DESIGN

12. Design a dc power supply to provide a +9V dc voltage as a replacement for a 9V battery. Be sure to check the data sheet of the voltage regulator selected for the required input dc voltage. (*Hint:* A different transformer may be required to increase the secondary voltage.) Build the circuit in Multisim to verify the design.

System Application 2

Lab Partner(s) _____

PARTS USED Nominal Value	Measured Value	PARTS USED Nominal Value	Measured Value
220 Ω		0.33 μF	
1 kΩ		1 μF	
10 kΩ		4.7 μF	
0.1 μF		10 μF	

	V_F	V_R	LM7805C PIN-OUT
1N4001			
1N4001			
1N4001			
1N4001			

TABLE SA2-1. VOLTAGE MEASUREMENTS WITH NO REGULATION   OBSERVATIONS

	Calculated	Measured	
		Multisim	Actual
	V_{TP-B}	V_{TP-B}	V_{TP-B}
DC load voltage			
Peak-to-peak ripple			

TABLE SA2-2. VOLTAGE MEASUREMENTS WITH A ZENER DIODE FOR REGULATION   OBSERVATIONS

	Measured			
	Multisim		Actual	
	V_{TP-B}	V_{TP-C}	V_{TP-B}	V_{TP-C}
DC load voltage				
Peak-to-peak ripple				

TABLE SA2-3. VOLTAGE MEASUREMENTS WITH A VOLTAGE REGULATOR   OBSERVATIONS

	Measured			
	Multisim		Actual	
	V_{TP-B}	V_{TP-C}	V_{TP-B}	V_{TP-C}
DC load voltage				
Peak-to-peak ripple				

	Measured			
	Multisim		Actual	
	$V_{TP\text{-}B}$	$V_{TP\text{-}C}$	$V_{TP\text{-}B}$	$V_{TP\text{-}C}$
$C_1 = 4.7\ \mu F$				
DC load voltage				
Peak-to-peak ripple				
$C_1 = 1\ \mu F$				
DC load voltage				
Peak-to-peak ripple				

DESIGN OBSERVATIONS

Questions for System Application 2

1. How would the voltage waveform across the load resistor be different had the capacitor been a value of $1\ \mu F$ instead of a $10\ \mu F$?

2. What is the purpose of R_1 in Fig. SA2-3?

3. How did the measured voltage waveform differ between points B and C on Fig. SA2-4?

4. Locate a data sheet for the LM7805. What is the purpose of the two capacitors (C_1 and C_2) shown in Fig. SA2-4?

5. The ac ripple on a dc voltage is best viewed with an oscilloscope that is _____ to the measurement point.
   (a) dc coupled;   (b) ac coupled;   (c) adjacent;   (d) transformer coupled.
6. How much of the peak voltage was dropped across the diodes in Fig. SA2-4 at any one time?
   (a) 1.4 V;   (b) 2.8 V;   (c) 0 V;   (d) 12 V.

TROUBLESHOOTING

7. As electrolytic capacitors age, their capacitance value _____.
   (a) intensifies;   (b) doubles in strength;   (c) becomes magnetized;
   (d) tends to decrease.
8. Care must be taken with electrolytic capacitors to observe proper polarity to avoid

   _____.
   (a) strong magnetic fields;   (b) electrical noise;   (c) destroying the capacitor;
   (d) static buildup.

DESIGN

9. Locate the data sheet for an LM7805C. What is the maximum current limit without the use of a heat sink?
   (a) 100 mA;   (b) 1 A;   (c) 5 V;   (d) 500 mA.
10. Locate the data sheet for an LM7805C. What is the minimum input voltage required to maintain voltage regulation?
    (a) 6 V;   (b) 5.7 V;   (c) 7.5 V;   (d) 8.2 V.

14

Optoelectronic Devices

In a forward-biased LED, heat and light are radiated when free electrons and holes recombine at the junction. Because the LED material is semitransparent, some of the light escapes to the surroundings. LEDs have a typical voltage drop from 1.5 and 2.5 V for currents between 10 and 50 mA. The exact voltage drop depends on the color, tolerance, and other factors. For troubleshooting and design, the second diode approximation with a knee of 2 V will be used.

An LED array is a group of LEDs that displays numbers, letters, or other symbols. The most common LED array is the seven-segment display. It contains seven rectangular LEDs. Each LED is called a segment because it forms one part of the character being displayed. By activating one or more LEDs, any digit from 0 through 9 can be formed.

An optocoupler combines an LED and a photodetector in a single package. As the light from the LED hits the photodetector, it produces an output voltage based on the amount of current flowing through the LED. If the LED current has an ac variation, V_{out} will have an ac variation. The key advantage of an optocoupler is the electrical isolation between the LED circuit and the output circuit, typically in thousands of megohms.

GOOD TO KNOW

The typical breakdown voltage for an LED is 3 to 5 V. It is important to insert the LED in the circuit correctly, or this voltage can easily be exceeded.

Required Reading

Chapter 5 (Sec. 5-8) of *Electronic Principles,* 8th ed.

Equipment

2   power supplies: one at 15 V, another adjustable from approximately 0 to 15 V
2   LEDs: L53RD and L53GD (or equivalent red and green LEDs)
2   1-W resistors: 270 Ω
8   ½-W resistors: 300 Ω
1   Seven-segment display: LDS-HTA514RI (or equivalent)
1   Optocoupler: 4N26 (or equivalent)
1   DMM (digital multimeter)

Procedure

DATA FOR A RED LED

1. Examine the red LED. Notice that one side of the package has a flat edge. This indicates the cathode side. (With many LEDs, the cathode lead is slightly shorter than the anode lead. This shorter lead is another way to identify the cathode.)
2. Build the circuit of Fig. 14-1 using a red LED.

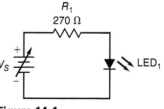

Figure 14-1

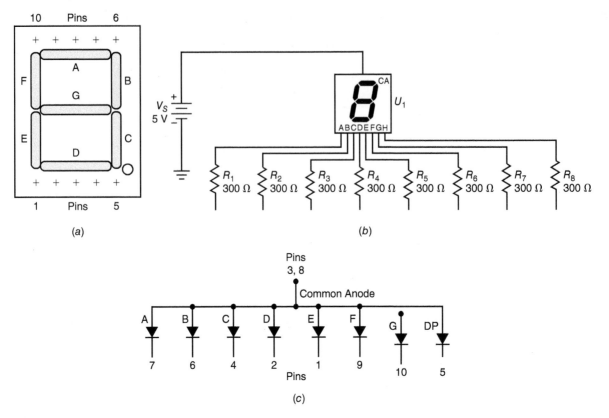

(a)

(b)

(c)

Figure 14-2

3. Adjust source voltage V_S to get 10 mA through the LED. Record the corresponding LED voltage in Table 14-1.
4. Adjust the source voltage and set up the remaining currents listed in Table 14-1. Record each LED voltage.

DATA FOR A GREEN LED

5. Replace the red LED with a green LED in the circuit of Fig. 14-1.
6. Repeat Steps 3 and 4 for the green LED.

USING A SEVEN-SEGMENT DISPLAY

7. Figure 14-2a shows the pinout for the seven-segment display used in this experiment (top view). It includes a right decimal point (RDP). Build the circuit of Fig. 14-2b.
8. Figure 14-2c shows the schematic diagram for a LDS-HTA514RI. Ground pins 4, 6, and 7. If the circuit is working correctly, digit 7 will be displayed.
9. Disconnect the grounds on pins 4, 6, and 7.
10. Refer to Fig. 14-2a and c. Which pins should be grounded to display a zero? Ground these pins and, if the circuit is working correctly, enter the pin numbers in Table 14-2.
11. Repeat Step 10 for the remaining digits, 1 through 9, and the decimal points.

THE TRANSFER GRAPH OF AN OPTOCOUPLER

12. Build the circuit of Fig. 14-3. Adjust the source voltage to 2 V. Measure and record the output voltage (Table 14-3).
13. Repeat Step 12 for the source voltages shown in Table 14-3. Record the corresponding output voltages.

TROUBLESHOOTING

14. If V_S is 15 V in Fig. 14-1, estimate the voltage across the red LED if it is open. Record the answer in Table 14-4. Similarly, estimate the LED voltage if the LED is shorted.
15. Build the circuit with a source voltage of 15 V and a red LED. Measure and record the LED voltage for the troubles listed in Table 14-4.

CRITICAL THINKING

16. Select a current-limiting resistor for a red LED in Fig. 14-1 that sets up a current of approximately 20 mA when the source voltage is 15 V. Record the resistor value in Table 14-5. Calculate and record the LED current and voltage.

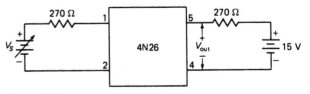

Figure 14-3

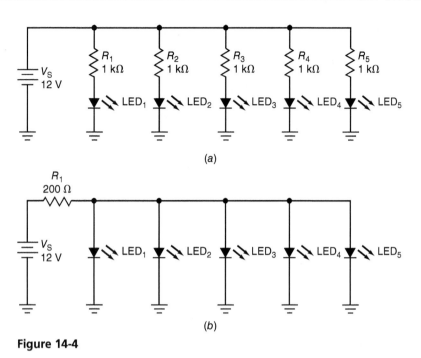

Figure 14-4

17. Build the design. Measure and record the LED current and voltage.
18. Repeat Steps 16 and 17 for the green LED.

APPLICATION (OPTIONAL)

19. Five red LEDs are going to be used to provide accent lighting under a cabinet. Assume each LED has a knee voltage of 2 V and requires 10 mA for maximum brightness. Which of the two circuit configurations shown in Fig. 14-4 would be a better design? Be sure to explain why it is a better choice.
20. Build the circuit selected in Step 19 and measure the current and voltage for each LED.
21. Turn off the power supply and remove one LED from the circuit to simulate a burnt-out LED. Measure the current and voltage for the remaining LEDs. Did the removal of the LED have any effect on the circuit?

Experiment 14

Lab Partner(s) _____

PARTS USED Nominal Value	Measured Value	PARTS USED Nominal Value	Measured Value
270 Ω		300 Ω	
270 Ω		300 Ω	
300 Ω		300 Ω	
300 Ω		300 Ω	
300 Ω		300 Ω	

TABLE 14-1. LED DATA

I	Multisim		Actual	
	V_{red}	V_{green}	V_{red}	V_{green}
10 mA				
20 mA				
30 mA				
40 mA				

TABLE 14-2. SEVEN-SEGMENT INDICATOR

Display	Multisim Pins Grounded	Actual Pins Grounded
0		
1		
2		
3		
4		
5		
6		
7		
8		
9		
10		
LDP		
RDP		

TABLE 14-3. OPTOCOUPLER

V_S	V_{out}
2 V	
4 V	
6 V	
8 V	
10 V	
12 V	
14 V	

TABLE 14-4. TROUBLESHOOTING

	Estimated V_{LED}	Measured V_{LED}	
		Multisim	Actual
Open LED			
Shorted LED			

TABLE 14-5. CRITICAL THINKING

		Calculated		Measured			
				Multisim		Actual	
	R	I_{LED}	V_{LED}	I_{LED}	V_{LED}	I_{LED}	V_{LED}
Red LED							
Green LED							

Questions for Experiment 14

1. The voltage drop across the red LED for a current of 30 mA was closest to: ( )
   (a) 0 V;   (b) 1 V;   (c) 2 V;   (d) 4 V.
2. The voltage drop across the green LED for a current of 30 mA was closest to: ( )
   (a) 0 V;   (b) 1 V;   (c) 2 V;   (d) 4 V.
3. To display a 1 on the seven-segment indicator, which pins did you ground? ( )
   (a) 1;   (b) 1 and 10;   (c) 10 and 13;   (d) 2, 7, and 8.
4. In Fig. 14-2, which of the following is true? ( )
   (a) LED brightness decreases as more segments are lit;   (b) all segments are equally bright at all times;   (c) number 8 was brighter than number 1.
5. When the source voltage increases in Fig. 14-3, the output voltage: ( )
   (a) decreases;   (b) stays the same;   (c) increases.
6. Explain how the LED array of Fig. 14-2 works. Include a discussion of the LED current versus the total current.

TROUBLESHOOTING

7. Why was the LED voltage large when the LED was open?

CRITICAL THINKING

8. Explain how the value of the current-limiting resistor for the red LED in Table 14-5 was calculated.

9. It is possible to get equal brightness of all numbers with the LED array of Fig. 14-2. How can this be done?

10. Optional. Instructor's question.

The CE Connection

As an approximation of transistor behavior, the Ebers-Moll model is used: The emitter diode acts like a rectifier diode, and the collector diode acts like a controlled-current source. The voltage across the emitter diode of a small-signal transistor is typically 0.6 to 0.7 V. For most troubleshooting and design, assume a 0.7 V for the V_{BE} drop. In this experiment, data will be gathered for calculating the α_{dc}, β_{dc}, and the V_{BE} drop.

When the maximum ratings of a transistor are exceeded, it can be damaged in several ways. The most common transistor trouble is a collector-emitter short where both the emitter diode and the collector diode are shorted. Another common transistor trouble is the collector-emitter open where both the emitter diode and the collector diode are open. Besides the foregoing, it is possible to have only one diode shorted, only one diode open, a leaky diode, etc.

To keep the troubleshooting straightforward, the two most common troubles will be emphasized: the collector-emitter short and the collector-emitter open. A collector-emitter short will be simulated by putting a jumper between the collector, base, and emitter; this shorts all three terminals together. The collector-emitter open is simulated by removing the transistor from the circuit; this opens both diodes.

GOOD TO KNOW

The pinout for the 2N3904 is shown to the right. The actual data sheet can be found on the Internet by searching for a 2N3904 data sheet.

Emitter Base Collector

Required Reading

Chapter 6 (Secs. 6-1 to 6-9) of *Electronic Principles*, 8th ed.

Equipment

1   power supply: 15 V
3   ½-W resistors: 100 Ω, 1 kΩ, 470 kΩ
3   transistors: 2N3904 (or almost any small-signal *npn* silicon transistor)
1   VOM (analog multimeter)
1   DMM (digital multimeter)

Procedure

TRANSISTOR TESTS

1. Using the VOM, measure the resistance between the collector and emitter of one of the transistors. This resistance should be extremely high (hundreds of megohms) in either direction.
2. Measure the forward and reverse resistance of the base-emitter diode and the collector-base diode. For both diodes, the reverse/forward resistance ratio should be at least 1000:1.
3. Repeat Steps 1 and 2 for two other transistors.

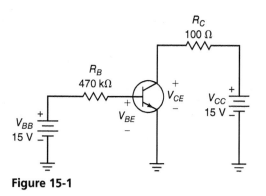

Figure 15-1

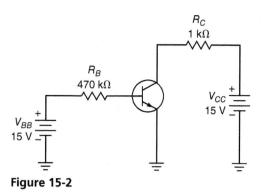

Figure 15-2

4. Test each of the transistors using the diode test range of the DMM. Measure and record the resistance value of the three resistors.

TRANSISTOR CHARACTERISTICS

5. Assume a V_{BE} of 0.7 V and β_{dc} = 200. Calculate and record I_B, I_C, I_E, V_{CB}, V_{CE}, and a_{dc} for Fig. 15-1 in Table 15-1. Build the circuit shown in Fig. 15-1 using one of the transistors.
6. Measure and record I_B, I_C, and I_E in Table 15-1.
7. Measure and record V_{BE} and V_{CE} in Table 15-1.
8. Calculate the values of α_{dc} and β_{dc} in Fig. 15-1. Record in Table 15-1.
9. Repeat Steps 5 to 8 using a second transistor. If using Multisim, only perform the measurement once.
10. Repeat Steps 5 to 8 using a third transistor.
11. If a curve tracer is available, display the collector curves of all three transistors. Notice the differences in β_{dc}, breakdown voltages, etc.

TROUBLESHOOTING

12. In Fig. 15-2, estimate and record the collector-to-ground voltage V_C for each trouble listed in Table 15-2. Note: To simulate a collector-emitter short, put a jumper between the collector, emitter, and base so that all three terminals are shorted together. To simulate a collector-emitter open, remove the transistor from the circuit.

13. Build the circuit with each of the foregoing troubles. Measure and record the collector voltage for each trouble.

CRITICAL THINKING

14. Build the circuit shown in Fig. 15-2 and measure the collector-to-ground voltage.
15. With the data from Step 13, calculate β_{dc} and select a base resistance that will produce a collector voltage of approximately half the supply voltage. Record the nearest standard resistance and the calculated collector voltage in Table 15-3.
16. Build the circuit with the design value for the base resistance. Complete the entries of Table 15-3.

ADDITIONAL WORK (OPTIONAL)

17. Repeat Steps 5 through 8 for a base resistance of 220 kΩ. Record all values on a separate piece of paper and compare these values to those recorded earlier.
18. Repeat Steps 5 through 8 for a base resistance of 100 kΩ. Record all values on a separate piece of paper and compare these values to those recorded earlier. What can be determined from the results?

Experiment 15

Lab Partner(s) _____

PARTS USED Nominal Value	Measured Value	PARTS USED	VOM Measured Value				DMM Measured Value			
			B-E		B-C		B-E		B-C	
			V_F	V_R	V_F	V_R	V_F	V_R	V_F	V_R
100 Ω		2N3904								
1 kΩ		2N3904								
470 kΩ		2N3904								

TABLE 15.1 TRANSISTOR VOLTAGES AND CURRENTS

	I_B	I_C	I_E	V_{BE}	V_{CE}	α_{dc}	β_{dc}
Calculated				0.7 V			200
Multisim							
Transistor							
1							
2							
3							

TABLE 15-2. TROUBLESHOOTING

Trouble	Estimated V_C	Measured V_C	
		Multisim	Actual
Open 470 kΩ			
Shorted 1 kΩ			
Open 1 kΩ			
Shorted collector-emitter			
Open collector-emitter			

TABLE 15-3. CRITICAL THINKING

	Calculated	Measured			
		Multisim	% Error	Actual	% Error
R_B					
V_C					

Questions for Experiment 15

1. The V_{BE} drop of the transistors was closest to: (   )
   (a) 0 V;   **(b)** 0.3 V;   **(c)** 0.7 V;   **(d)** 1 V.
2. The α_{dc} of all transistors was very close to: (   )
   (a) 0;   **(b)** 1;   **(c)** 5;   **(d)** 20.
3. The β_{dc} of all transistors was greater than: (   )
   (a) 0;   **(b)** 1;   **(c)** 5;   **(d)** 20.
4. This experiment proves that collector current is much greater than: (   )
   (a) collector voltage;   **(b)** emitter current;   **(c)** base current;   **(d)** 0.7 V.
5. The transistors were silicon because: (   )
   (a) V_{BE} was approximately 0.7 V;   **(b)** I_C is much greater than I_B;   **(c)** the collector diode was reverse biased;   **(d)** β_{dc} was much greater than unity.
6. What was the typical V_{BE} drop and how did the collector current relate to the base current?

TROUBLESHOOTING

7. What collector voltage was measured when the base resistor was open? Explain why this voltage existed at the collector.

8. Briefly explain why the collector voltage is approximately zero when a transistor has a collector-emitter short.

CRITICAL THINKING

9. Explain how the base resistance was calculated in Table 15-3.

10. Why is it that increasing or decreasing V_{CC} or R_C in Fig. 15-1 does not change the collector current?

Transistor Operating Regions

To determine the region of transistor operation (cutoff, saturation, or active), two questions must be asked: First, is there base current? Second, is there collector current? To have base current, two conditions are necessary: A complete path must exist for base current, and a voltage must be applied somewhere in this path. Similar conditions apply to the collector circuit.

When there is no base current, the transistor goes into cutoff. For instance, if the base resistor of Fig. 16-1 were open, the path for base current would be broken. In this case, base current would be zero. Since the collector current equals the dc beta times the base current, the collector current would be ideally zero and the transistor would be cut off. In a circuit like Fig. 16-1, no collector current implies that $V_{CE} = V_{CC} = 10$ V.

When there is base current, V_{BE} is approximately 0.7 V. In this case, the transistor may operate in any of the three regions: active, saturation, or cutoff. As a guide, a small-signal transistor is operating in the active region when the collector-emitter voltage V_{CE} is greater than 1 V but less than V_{CC}. Saturation occurs somewhere below 1 V, depending on the transistor type. For a typical small-signal transistor, $V_{CE(sat)}$ is 0.1 to 0.2 V.

In this experiment, the voltages present in various transistor circuit configurations will be calculated and measured. Based on the data collected, it will be possible to determine whether the transistor operates in the active, saturation, or cutoff region.

GOOD TO KNOW

The pinout for the 2N3904 is shown to the right. The actual data sheet can be found on the Internet by searching for a 2N3904 data sheet.

Emitter Base Collector

Required Reading

Chapter 6 of *Electronic Principles*, 8th ed.

Equipment

1 power supply: adjustable to 10 V
1 DMM (digital multimeter)
1 transistor: 2N3904
3 ½-W resistors: 1 kΩ, 10 kΩ, 470 kΩ

Procedure

1. Using a DMM in the diode test mode, measure and record the forward- and reverse-biased voltages. When forward biased, the DMM should read approximately 0.7 V. When reverse biased, the DMM should read OL. Assume a dc beta of 200 in Fig. 16-1. Calculate V_{BE} and V_{CE}. In which of the three regions is the transistor operating? Record the voltages and the region in Table 16-1.

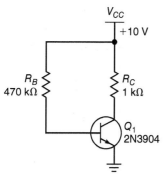

Figure 16-1

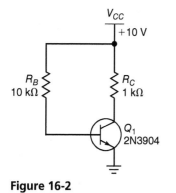

Figure 16-2

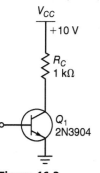

Figure 16-3

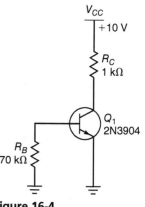

Figure 16-4

> **GOOD TO KNOW**
>
> When operating in saturation, the transistor acts like a closed switch between the collector and emitter. When operating in cutoff, the transistor acts like an open switch between the collector and emitter.

2. Build the circuit shown in Fig. 16-1. Measure V_{BE} and V_{CE}. In which of the three regions is the transistor operating? Record the voltages and the region in Table 16-1.
3. Repeat Steps 1 and 2 for Fig. 16-2. Use Table 16-2.
4. Repeat Steps 1 and 2 for Fig. 16-3. Use Table 16-3.
5. Repeat Steps 1 and 2 for Fig. 16-4. Use Table 16-4.
6. Figure 16-5 shows a circuit with four transistors, labeled Q_1 to Q_4. Estimate the values of V_{BE} and V_{CE} for each transistor, and then determine the operating region for the transistor. Record the estimated voltages and the operating regions in Table 16-5.

TROUBLESHOOTING

7. Suppose the base resistor is open in Fig. 16-1. What are the values of V_{BE} and V_{CE}? In what region is the transistor operating? Record the answers in Table 16-6 next to "R_{BO}."
8. In attempting to connect the circuit of Fig. 16-1, the base resistor is accidently connected to ground instead of the supply voltage. This results in the circuit of Fig. 16-4. What are the values of V_{BE} and V_{CE}? In what region is the transistor operating? Record the answers in Table 16-6 next to "R_{BG}."
9. Suppose the collector resistor is open in Fig. 16-1. What are the values of V_{BE} and V_{CE}? In what region is the transistor operating? Record the answers in Table 16-6 next to "R_{CO}."
10. Suppose the collector resistor is shorted in Fig. 16-1. What are the values of V_{BE} and V_{CE}? In what region is the transistor operating? Record the answers in Table 16-6 next to "R_{CS}."
11. Build the circuit shown in Fig. 16-1 with a shorted collector resistance. Then, measure V_{BE} and V_{CE} to verify reasonable agreement with the values recorded for R_{CS} in Table 16-6.
12. Build the circuit shown in Fig. 16-1 with an open collector resistance. Then, measure V_{BE} and V_{CE} to verify reasonable agreement with the values recorded for R_{CO} in Table 16-6.

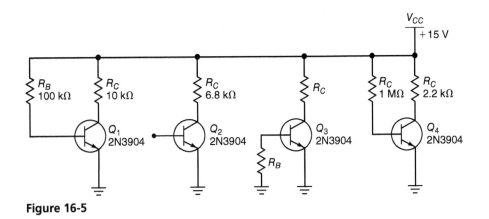

Figure 16-5

CRITICAL THINKING

13. Based on the data collected in Table 16-1, determine the value of R_B required, to get $V_{CE} = 2$ V.
14. Round off the calculated resistance to the nearest standard size. Then, build the circuit of Fig. 16-1 with the foregoing resistance. Measure V_{CE} to verify that it is reasonably close to 2 V.

ADDITIONAL WORK (OPTIONAL)

15. Based on calculated results, determine the region the transistor in Fig. 16-1 operates in.

16. Based on calculated results, determine the region the transistor in Fig. 16-2 operates in.

17. Based on calculated results, determine the region the transistor in Fig. 16-3 operates in.

18. Based on calculated results, determine the region the transistor in Fig. 16-4 operates in.

Experiment 16

Lab Partner(s) _____

PARTS USED Nominal Value	Measured Value	PARTS USED	Measured Value			
			B-E		B-C	
1 kΩ			V_F	V_R	V_F	V_R
10 kΩ						
470 kΩ		2N3904				

TABLE 16-1. FIRST CIRCUIT

	Calculated	Measured	
		Multisim	Actual
V_{BE}			
V_{CE}			
Region			

TABLE 16-2. SECOND CIRCUIT

	Calculated	Measured	
		Multisim	Actual
V_{BE}			
V_{CE}			
Region			

TABLE 16-3. THIRD CIRCUIT

	Calculated	Measured	
		Multisim	Actual
V_{BE}			
V_{CE}			
Region			

TABLE 16-4. FOURTH CIRCUIT

	Calculated	Measured	
		Multisim	Actual
V_{BE}			
V_{CE}			
Region			

TABLE 16-5. FIFTH CIRCUIT

	Q_1	Q_2	Q_3	Q_4
V_{BE}				
V_{CE}				
Region				

TABLE 16-6. TROUBLESHOOTING

	Calculated		Measured				Region
			Multisim		Actual		
	V_{BE}	V_{CE}	V_{BE}	V_{CE}	V_{BE}	V_{CE}	
R_{BO}							
R_{BG}							
R_{CO}							
R_{CS}							

Questions for Experiment 16

1. The base current in Fig. 16-1 is closest to: ( )
   (a) 0;   (b) 20 μA;   (c) 470 μA;   (d) 10 mA.
2. The collector current in Fig. 16-2 is closest to: ( )
   (a) 0;   (b) 20 μA;   (c) 470 μA;   (d) 10 mA.
3. The collector current in Fig. 16-3 is closest to: ( )
   (a) 0;   (b) 20 μA;   (c) 470 μA;   (d) 10 mA.
4. The collector voltage in Fig. 16-4 is closest to: ( )
   (a) 0;   (b) 2 V;   (c) 5 V;   (d) 10 V.
5. In Fig. 16-5, Q_1 operates in which region?   ( )
   (a) active;   (b) saturation;   (c) cutoff;   (d) breakdown.
6. In Fig. 16-5, Q_4 operates in which region?   ( )
   (a) active;   (b) saturation;   (c) cutoff;   (d) breakdown.
7. When the collector resistor of Fig. 16-1 is shorted, V_{CE} equals: ( )
   (a) 0;   (b) 2 V;   (c) 5 V;   (d) 10 V.
8. When the collector resistor of Fig. 16-1 is open, V_{CE} is closest to: ( )
   (a) 0;   (b) 2 V;   (c) 5 V;   (d) 10 V.
9. A technician is troubleshooting a circuit like Fig. 16-1 and has measured the collector-emitter voltage. How can the region the transistor is operating in be determined?

10. Optional. Instructor's question.

Base Bias

A circuit like Fig. 17-1 is referred to as base bias because it sets up a fixed base current. The base current can be calculated by applying Ohm's law to the total base resistance. This base current will remain constant when the transistors are replaced.

On the other hand, the collector current equals the current gain times the base current. Because of this, the collector current may have large variations from one transistor to the next. In other words, the Q point in a base-biased circuit is heavily dependent on the value of β_{dc}.

GOOD TO KNOW

The pinout for the 2N3904 is shown to the right. The actual data sheet can be found on the Internet. Search for the 2N3904 data sheet.

Emitter Base Collector

Required Reading

Chapter 6 (Secs. 6-10 to 6-15) of *Electronic Principles,* 8th ed.

Equipment

1 power supply: 15 V
3 transistors: 2N3904 (or almost any small-signal *npn* silicon transistor)
2 ½-W resistors: 2.2 kΩ and 22 kΩ
1 decade resistance box (or substitute a 1-MΩ potentiometer)

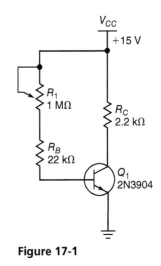

Figure 17-1

Procedure

1. Measure and record the value of each resistor. The fixed-base-current circuit of Fig. 17-1 is not a stable biasing circuit, but it is a good way to measure β_{dc}.
2. Build the circuit shown in Fig. 17-1 using one of the three transistors.
3. Adjust R_1 to get a V_{CE} of 1 V. Record the value of R_1 in Table 17-1. (If R is a potentiometer instead of a

decade box, it will have to be removed from the circuit in order to measure its resistance.) In Fig. 17-1, notice the total base resistance R_B equals R_1 plus 22 kΩ. Record the value of R_B in Table 17-1.
4. Calculate the values of β_{dc} and I_C. Record in Table 17-1.
5. Repeat Steps 2 through 4 for the second and third transistors. If Multisim is being used, change the β_{dc} of transistors 2 and 3 to 150 and 100, respectively.

6. With the values of Table 17-1, calculate the ideal and second-approximation values of I_E in Fig. 17-1. Record the I_E values in Table 17-2.

7. If a curve tracer or other transistor tester is available, measure the β_{dc} of each transistor for an I_C of approximately 5 mA. The values should be similar to the β_{dc} values of Table 17-1.

IDENTIFYING THE OPERATING REGION

8. In Fig. 17-1, calculate V_{BE}, $I_{C(sat)}$, $V_{CE(sat)}$, and $V_{CE(cutoff)}$. Record these values in Table 17-3.

9. Adjust R_1 to 0 Ω and measure V_{BE}, I_C, and V_{CE}. Record these values in Table 17-4. Compare the measured values to the calculated values of Table 17-3. Decide which region the transistor is operating in. Record the answer in Table 17-4.

10. Disconnect one end of the 22 kΩ base resistor and measure V_{BE}, I_C, and V_{CE}. Record these values in Table 17-4. Decide which region the transistor is now operating in. Record the answer.

11. Adjust R_1 until V_{CE} is approximately 7.5 V. Measure V_{BE} and I_C. Record these values and the actual value of V_{CE} in Table 17-4. Which region is the transistor operating in? Record the answer.

APPLICATION (OPTIONAL)

12. A photocell can control a base-biased circuit to turn on different loads. As a demonstration, build the circuit of Fig. 17-2. The photocell should be exposed to incoming light.

13. Adjust R_1 until the buzzer begins to buzz and the LED lights up. Readjust R_1 slowly until the buzzer and LED just turn off.

14. Now, cover the photocell to prevent light from reaching it. The buzzer and LED should come on.

15. Repeat Steps 13 and 14 for a few other photocells.

ADDITIONAL WORK (OPTIONAL)

16. In this section, the measured values of the base and collector currents will be used to create a graph of β_{dc} versus I_C.

17. Build the circuit of Fig. 17-3. Measure I_B and I_C for the minimum value of R_1. Record the values of I_B and I_C on a separate piece of paper.

18. Measure and record the foregoing currents for the maximum value of R_1.

19. Measure and record the foregoing currents for several intermediate values of R_1.

20. Calculate dc beta for each value of I_C. Then, graph dc beta versus the collector current. If additional data points are needed, repeat Step 19.

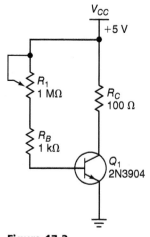

Figure 17-3

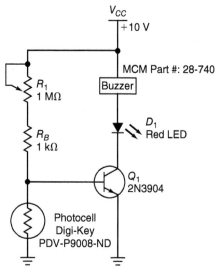

Figure 17-2

Experiment 17

Lab Partner(s) _____

PARTS USED Nominal Value	Measured Value
2.2 kΩ	
22 kΩ	
1 MΩ	

CALCULATIONS

TABLE 17-1. β_{dc} VALUES

Transistor	Multisim				Actual			
	R_1	R_B	β_{dc}	I_C	R_1	R_B	β_{dc}	I_C
1								
2								
3								

TABLE 17-2. CALCULATIONS

Test	Multisim		Actual	
	$I_{E(ideal)}$	$I_{E(second)}$	$I_{E(ideal)}$	$I_{E(second)}$
1				
2				
3				

TABLE 17-3. CALCULATIONS

Multisim				Actual			
V_{BE}	$I_{C(sat)}$	$V_{CE(sat)}$	$V_{CE(cutoff)}$	V_{BE}	$I_{C(sat)}$	$V_{CE(sat)}$	$V_{CE(cutoff)}$

TABLE 17-4. MEASUREMENTS

Condition	Multisim				Actual			
	V_{BE}	I_C	V_{CE}	Operating region	V_{BE}	I_C	V_{CE}	Operating region
$R_1 = 0$								
22 kΩ open								
$V_{CE} \approx 7.5$ V								

Questions for Experiment 17

1. In Fig. 17-1, an increase in current gain causes an increase in: ( )
   (a) I_C;   (b) V_{CE};   (c) I_B;   (d) V_{CC}.
2. When R_B increases in a base-biased circuit, which of these increases? ( )
   (a) I_E;   (b) V_{BB};   (c) I_C;   (d) V_{CE}.
3. When R_C increases in a base-biased circuit, which of these decreases? ( )
   (a) I_E;   (b) V_{BB};   (c) I_C;   (d) V_{CE}.
4. When V_{BB} increases in a base-biased circuit, which of these decreases? ( )
   (a) I_B;   (b) V_{BE};   (c) I_C;   (d) V_{CE}.
5. When β_{dc} increases in a base-biased circuit, which of these remains constant? ( )
   (a) I_B;   (b) I_E;   (c) I_C;   (d) V_{CE}.
6. The ideal and second-approximation values in Table 17-2 differ by approximately: ( )
   (a) 0.1%;   (b) 1%;   (c) 5%;   (d) 10%.
7. If the base current is 10 μA in Fig. 17-1 and the collector voltage is 10 V, the current ( )
   gain is closest to:
   (a) 50;   (b) 125;   (c) 225;   (d) 350.
8. If the collector voltage is 5 V in Fig. 17-1 and β_{dc} is 150, the base current is ( )
   closest to:
   (a) 10 μA;   (b) 20 μA;   (c) 30 μA;   (d) μA.
9. Explain how a technician can identify when a transistor is operating in the cutoff region.

10. Explain how a technician can identify when a transistor is operating in the saturation region.

11. Explain how a technician can identify when a transistor is operating in the active region.

LED Drivers

The simplest way to use a transistor is as a switch, meaning that it operates at either saturation or cutoff but nowhere else along the load line. When saturated, a transistor appears as a closed switch between its collector and emitter terminals. When cut off, it is like an open switch. Because of the wide variation in β_{dc}, hard saturation is used with transistor switches. This means having enough base current to guarantee transistor saturation under all operating conditions. With small-signal transistors, hard saturation requires a base current of approximately one-tenth of the collector saturation current.

Another basic way to use the transistor is as a current source. In this case, the base resistor is omitted and the base supply voltage is connected directly to the base terminal. An emitter resistor is used to set up the desired collector current. The emitter voltage is V_{BE} volts lower than the base voltage V_{BB} ($V_{RE} = V_{BB} - V_{BE}$). Therefore, the collector current equals ($V_{BB} - V_{BE}$) divided by R_E. This fixed collector current then flows through the load, which is connected between the collector and positive supply voltage.

In this experiment, a transistor switch and a transistor current source will be constructed, along with an opportunity to troubleshoot and design these basic transistor circuits.

GOOD TO KNOW

The typical breakdown voltage for an LED is 3 to 5 V. It is important to insert the LED in the circuit correctly, or this voltage can easily be exceeded. The cathode will have a shorter leg and a flat side on the case.

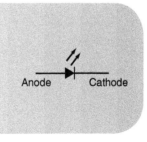

Required Reading

Chapter 6 (Secs. 6-10 to 6-16) of *Electronic Principles*, 8th ed.

Equipment

2 power supplies: each source adjustable from 0 to 15 V
3 ½-W resistors: 220 Ω, 1 kΩ, 10 kΩ
1 LED: L53RD (or an equivalent red LED)
3 transistors: 2N3904
1 DMM (digital multimeter)

Procedure

1. Measure and record the resistor values. In Fig. 18-1, calculate I_B, I_C, and V_{CE}. Record the answers in Table 18-1.
2. Build the transistor switch of Fig. 18-1. Measure and record the quantities listed in Table 18-1.

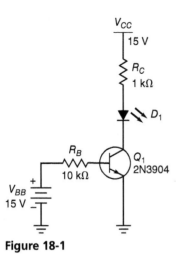

Figure 18-1

3. Repeat Steps 1 and 2 for the other transistors. If Multisim is being used, change the β_{dc} of transistors 2 and 3 to 150 and 100, respectively.

TRANSISTOR CURRENT SOURCE

4. In Fig. 18-2, calculate all quantities listed in Table 18-2.
5. Build the transistor current source of Fig. 18-2. Measure and record the quantities listed in Table 18-2.
6. Repeat Steps 4 and 5 for the other transistors.

TROUBLESHOOTING

7. In Fig. 18-1, assume the base resistor is open. Estimate and record the collector voltage in Table 18-3.
8. Repeat Step 7 for each of the troubles listed in Table 18-3.
9. Build the circuit of Fig. 18-1 with each of the troubles listed in Table 18-3. Measure and record all listed quantities.

10. In Fig. 18-2, assume the emitter resistor is open. Estimate and record the voltages listed in Table 18-4.
11. Repeat Step 10 for each of the troubles listed in Table 18-4.
12. Build the circuit of Fig. 18-2 with each of the troubles listed in Table 18-4. Measure and record all listed quantities.

CRITICAL THINKING

13. Select a collector resistance in Fig. 18-1 to produce a collector current of approximately 30 mA. Calculate and record the quantities listed in Table 18-5.
14. Build the circuit of Fig. 18-1 with the calculated value of collector resistance. Measure and record the quantities of Table 18-5.
15. Select an emitter resistance in Fig. 18-2 to get a collector current of approximately 30 mA. Calculate and record the quantities listed in Table 18-5.
16. Build the circuit of Fig. 18-2 with the calculated value of emitter resistance. Measure and record all quantities in Table 18-5.

APPLICATION (OPTIONAL)

17. Measure the light and dark resistance of a few cadmium-sulfide photocells such as the Digikey part #: PDV-P9008-ND. The photocell is a photo-resistive device, whose resistance changes with the amount of incoming light.
18. Build the circuit of Fig. 18-3. Adjust the supply voltage until the LED is dimly lit. Cover the photocell and notice how the LED either goes out or becomes less bright. Repeat this demonstration with a few other photocells.

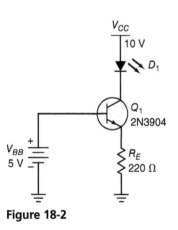

Figure 18-2

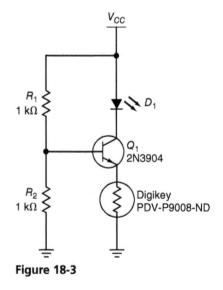

Figure 18-3

ADDITIONAL WORK (OPTIONAL)

19. Build the circuit of Fig. 18-4. Record whether the LED is on or off for each position of the switch.

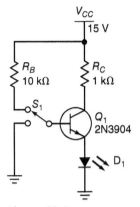

Figure 18-4

20. Build the circuit of Fig. 18-5. For each switch position, compare the on-off state of the LED to the recorded values of Step 19. What was learned?

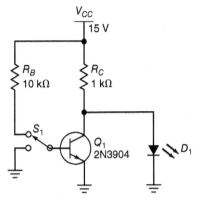

Figure 18-5

Experiment 18

Lab Partner(s) _____

PARTS USED Nominal Value	Measured Value
220 Ω	
1 kΩ	
10 kΩ	

CALCULATIONS

TABLE 18-1. TRANSISTOR SWITCH

	Calculated			Measured					
				Multisim			Actual		
Transistor	I_B	I_C	V_{CE}	I_B	I_C	V_{CE}	I_B	I_C	V_{CE}
1									
2									
3									

TABLE 18-2. TRANSISTOR CURRENT SOURCE

	Calculated			Measured					
				Multisim			Actual		
Transistor	I_E	I_C	V_{CE}	I_E	I_C	V_{CE}	I_E	I_C	V_{CE}
1									
2									
3									

TABLE 18-3. TROUBLESHOOTING THE TRANSISTOR SWITCH

		Measured V_C	
Trouble	Estimated V_C	Multisim	Actual
Open 10 kΩ			
Open 1 kΩ			
Shorted C-E			
Open C-E			

TABLE 18-4. TROUBLESHOOTING THE TRANSISTOR CURRENT SOURCE

| | Estimated | | Measured | | | |
| | | | Multisim | | Actual | |
Trouble	V_C	V_E	V_C	V_E	V_C	V_E
Open 220 Ω						
Shorted C-E						
Open C-E						

TABLE 18-5. CRITICAL THINKING

| | Calculated | | | Measured | | | | | |
| | | | | Multisim | | | Actual | | |
Transistor	R	V_E	I_C	V_E	I_C	% Error	V_E	I_C	% Error
Switch									
Current Source									

Questions for Experiment 18

1. In Fig. 18-1, the ratio of collector current to base current is closest to: ( )
   (a) 1;   (b) 10;   (c) 100;   (d) 300.
2. The measured V_{CE} entries of Table 18-1 indicate that collector voltage is approximately: ( )
   (a) 0;   (b) 2 V;   (c) 4 V;   (d) 8 V.
3. In the transistor current source of Fig. 18-2, the emitter voltage is closest to: ( )
   (a) 0.7 V;   (b) 4.3 V;   (c) 5 V;   (d) 10 V.
4. When a transistor is in hard saturation, its collector-emitter terminals appear approximately: ( )
   (a) shorted;   (b) open;   (c) in the active region;   (d) cut off.
5. With a transistor current source, the emitter voltage is V_{BE} volts lower than the: ( )
   (a) base voltage;   (b) emitter voltage;   (c) collector voltage;   (d) collector current.
6. What are some of the differences between a transistor switch and a transistor current source?

TROUBLESHOOTING

7. While troubleshooting a transistor switch like Fig. 18-1, the collector voltage is measured to be zero. If the LED is lit, what is the most likely trouble?

8. Explain the measured values for collector and emitter voltage when the emitter resistor was open in Fig. 18-2.

9. Why is hard saturation used with a transistor switch?

10. Optional. Instructor's question.

Setting Up a Stable *Q* Point

To achieve a stable *Q* point, either voltage-divider bias or two-supply emitter bias is required. With either of these stable biasing methods, the effects of h_{FE} variations are virtually eliminated. Voltage-divider bias requires only a single power supply. This type of bias is also called universal bias, an indication of its popularity. When two supplies are available, two-supply emitter bias can provide as stable a *Q* point as voltage-divider bias.

In this experiment, circuits utilizing both types of bias will be constructed and the stability of their *Q* points will be verified.

GOOD TO KNOW

The dc beta (β_{dc}) is listed as h_{FE} on the transistor data sheet.

Emitter Base Collector

Required Reading

Chapter 7 (Secs. 7-5 to 7-8) of *Electronic Principles*, 8th ed.

Equipment

2 power supplies: 15 V
3 transistors: 2N3904 (or equivalent)
5 ½-W resistors: 1 kΩ, 2.2 kΩ, 3.9 kΩ, 8.2 kΩ, 10 kΩ
1 DMM (digital multimeter)

Procedure

VOLTAGE-DIVIDER BIAS

1. Measure and record the values of the resistors. In Fig. 19-1, calculate V_B, V_E, and V_C. Record the answers in Table 19-1.
2. Build the circuit of Fig. 19-1. Measure and record the quantities listed in Table 19-1.
3. Repeat Steps 1 and 2 for the other transistors. If Multisim is being used, change the β_{dc} of transistors 2 and 3 to 250 and 150, respectively.

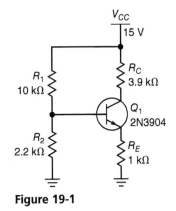

Figure 19-1

EMITTER BIAS

4. In Fig. 19-2, calculate V_B, V_E, and V_C. Record the answers in Table 19-2.
5. Build the emitter-biased circuit of Fig. 19-2. Measure and record the quantities of Table 19-2.
6. Repeat Steps 4 and 5 for the other transistors.

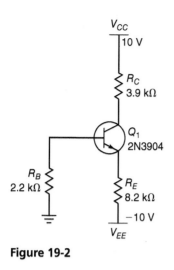

Figure 19-2

TROUBLESHOOTING

7. In Fig. 19-1, assume that R_1 is open. Estimate and record the collector voltage V_C in Table 19-3.
8. Repeat Step 7 for the other troubles listed in Table 19-3. Build the circuit of Fig. 19-1 with each trouble listed in Table 19-3. Measure and record the collector voltage.

CRITICAL THINKING

9. Design a stiff voltage-divider biased circuit to meet the following specifications: $V_{CC} = 15$ V, $I_C = 2$ mA, and $V_C = 7.5$ V. Assume an h_{FE} of 200. Calculate and record the quantities listed in Table 19-4.
10. Build the design. Measure and record the quantities of Table 19-4.

ADDITIONAL WORK (OPTIONAL)

11. Assume $\beta_{dc} = 172$ for the emitter-feedback biased circuit of Fig. 19-3. Calculate and record V_B, V_E, and V_C on a separate piece of paper. (Use a table similar to Table 19-1 for your data.)
12. Build the circuit of Fig. 19-3. Measure and record V_B, V_E, and V_C.

13. Compare the measured values to the calculated values. What does this say about β_{dc}?
14. Repeat Steps 11 to 13 for the collector-feedback biased circuit of Fig. 19-4.

GOOD TO KNOW

Since $I_C = \beta I_B$ and $I_E = I_C + I_B$, βI_B can be substituted for I_C.

$$I_E = \beta I_B + I_B = I_B(\beta+1)$$

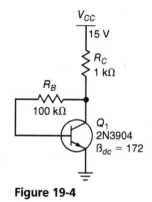

Figure 19-4

Experiment 19

Lab Partner(s) _____

PARTS USED Nominal Value	Measured Value	PARTS USED Nominal Value	Measured Value
1 kΩ		8.2 kΩ	
2.2 kΩ		10 kΩ	
3.9 kΩ			

CALCULATIONS

TABLE 19-1. VOLTAGE-DIVIDER BIAS

	Calculated			Measured					
				Multisim			Actual		
Transistor	V_B	V_E	V_C	V_B	V_E	V_C	V_B	V_E	V_C
1									
2									
3									

TABLE 19-2. EMITTER BIAS

	Calculated			Measured					
				Multisim			Actual		
Transistor	V_B	V_E	V_C	V_B	V_E	V_C	V_B	V_E	V_C
1									
2									
3									

TABLE 19-3. TROUBLESHOOTING

Trouble	Estimated V_C	Measured V_C	
		Multisim	Actual
Open R_1			
Shorted R_1			
Open R_2			
Shorted R_2			
Open R_C			
Shorted R_C			
Open R_E			
Shorted R_E			
Open C-E			
Shorted C-E			

TABLE 19-4. CRITICAL THINKING

Values: $R_1 =$ ____ ; $R_2 =$ ____ ; $R_C =$ ____ ; $R_E =$ ____					
			Measured V_C		
Transistor	Calculated V_C	Multisim	% Error	Actual	% Error
1					
2					
3					

Questions for Experiment 19

1. Ideally, the voltage divider of Fig. 19-1 produces which of the following base ( )
   voltages:
   (a) 0 V; **(b)** 1.1 V; **(c)** 1.8 V; **(d)** 6.03 V.
2. The measured emitter voltage of Fig. 19-1 was closest to: ( )
   (a) 0 V; **(b)** 1.1 V; **(c)** 1.8 V; **(d)** 6.03 V.
3. The measured collector voltage of Fig. 19-1 was closest to: ( )
   (a) 0 V; **(b)** 1.1 V; **(c)** 1.8 V; **(d)** 6.03 V.
4. The base voltage measured in Fig. 19-2 was: ( )
   (a) 0 V; **(b)** slightly positive; **(c)** slightly negative; **(d)** −0.7 V.
5. With both voltage-divider bias and emitter bias, the measured collector voltage was ( )
   approximately:
   (a) constant; **(b)** negative; **(c)** unstable; **(d)** one V_{BE} drop less than
   the base voltage.
6. What was discovered about the Q point of a circuit that uses voltage-divider bias or emitter
   bias?

TROUBLESHOOTING

7. Name all the troubles that were found to produce a collector voltage of 10 V.

8. What collector voltage was measured with a shorted collector-emitter? Explain why this
   value occurred.

CRITICAL THINKING

9. Compare the measured V_C with the calculated V_C in Table 19-4. Explain why the measured
   and calculated values differ.

10. Optional. Instructor's question.

Experiment 20

Biasing PNP Transistors

Since the emitter and collector diodes of a *pnp* transistor point in the opposite direction of an *npn* transistor, all currents and voltages are reversed in the *pnp* transistor. If only positive power supplies are available, the *pnp* transistor is connected upside down. In this experiment, *pnp* biasing circuits that work with positive or negative supply voltages will be built and analyzed.

GOOD TO KNOW

The pinout for the 2N3906 is shown to the right. The actual data sheet can be found on the Internet. Search for the 2N3906 data sheet.

Emitter Base Collector

Required Reading

Chapter 7 (Sec. 7-11) of *Electronic Principles,* 8th ed.

Equipment

1 power supply: ±15 V
3 *pnp* transistors: 2N3906 (or equivalent)
4 ½-W resistors: 1 kΩ, 2.2 kΩ, 3.9 kΩ, 10 kΩ
1 DMM (digital multimeter)

Procedure

NEGATIVE POWER SUPPLY

1. Measure and record the values of the resistors. In Fig. 20-1, calculate V_B, V_E, and V_C. Record the answers in Table 20-1.
2. Build the circuit of Fig. 20-1. Measure and record the quantities listed in Table 20-1.
3. Repeat Steps 1 and 2 for the other transistors. If Multisim is being used, change the β_{dc} of transistors 2 and 3 to 250 and 150, respectively.

POSITIVE POWER SUPPLY

4. In Fig. 20-2, calculate V_B, V_E, and V_C. Record the answers in Table 20-2.

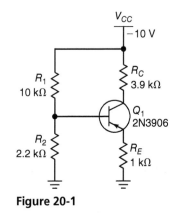

Figure 20-1

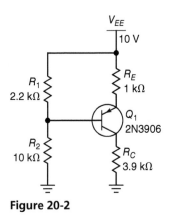

Figure 20-2

5. Build the circuit of Fig. 20-2. Measure and record the quantities of Table 20-2.
6. Repeat Steps 4 and 5 for the other transistors.

TROUBLESHOOTING

7. In Fig. 20-2, assume that R_1 is open. Estimate and record all voltages listed in Table 20-3.
8. Repeat Step 7 for the other troubles listed in Table 20-3.
9. Build the circuit of Fig. 20-2 with each trouble listed in Table 20-3. Measure and record all voltages.
10. Ask the instructor or another student to insert a trouble in the circuit.
11. Locate the trouble logically as follows. Measure V_B, V_E, and V_C. Then look at Fig. 20-2, visualizing these voltages. Try to figure out which trouble would produce these voltages. When the suspected trouble has been located, confirm it by consulting Table 20-3.
12. Repair the trouble and check that the circuit is working correctly.
13. Repeat Steps 10 to 12 as often as the instructor indicates.

CRITICAL THINKING

14. Design an LED driver like Fig. 20-3 with a stiff voltage divider to meet the following specifications: $V_{CC} = 5$ V and $I_C = 20$ mA. A typical h_{FE} of 200 may be assumed. Record the design values (nearest standard resistances) in Table 20-4. Calculate and record I_C for the design.
15. Build the design. Measure and record I_C. Repeat this measurement for the other transistors.

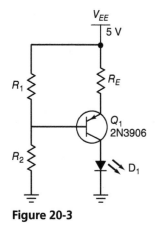

Figure 20-3

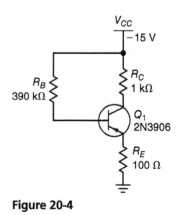

Figure 20-4

ADDITIONAL WORK (OPTIONAL)

16. Assume $\beta_{dc} = 100$ for the emitter-feedback biased circuit of Fig. 20-4. Calculate and record V_B, V_E, and V_C on a separate piece of paper. (Use a table similar to Table 20-1 for the data.)
17. Build the circuit of Fig. 20-4. Measure and record V_B, V_E, and V_C.
18. Compare the measured values to the calculated values. What does this say about β_{dc}?
19. Repeat Steps 16 to 18 for the collector-feedback biased circuit of Fig. 20-5.

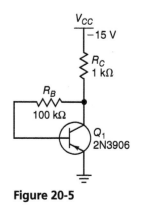

Figure 20-5

Experiment 20

Lab Partner(s) _____

PARTS USED Nominal Value	Measured Value
1 kΩ	
2.2 kΩ	
3.9 kΩ	
10 kΩ	

CALCULATIONS

TABLE 20-1. NEGATIVE POWER SUPPLY

	Calculated			Measured					
				Multisim			Actual		
Transistor	V_B	V_E	V_C	V_B	V_E	V_C	V_B	V_E	V_C
1									
2									
3									

TABLE 20-2. EMITTER BIAS

	Calculated			Measured					
				Multisim			Actual		
Transistor	V_B	V_E	V_C	V_B	V_E	V_C	V_B	V_E	V_C
1									
2									
3									

TABLE 20-3. TROUBLESHOOTING

	Estimated			Measured					
				Multisim			Actual		
Trouble	V_B	V_E	V_C	V_B	V_E	V_C	V_B	V_E	V_C
Open R_1									
Open R_2									
Shorted R_2									
Open R_C									
Shorted R_C									
Open R_E									
Shorted R_E									
Open C-E									
Shorted C-E									

TABLE 20-4. CRITICAL THINKING

		Measured I_C			
Transistor	Calculated I_C	Multisim	% Error	Actual	% Error
1					
2					
3					

Values: $R_1 =$ _____; $R_2 =$ _____; $R_E =$ _____

Questions for Experiment 20

1. Ideally, the voltage divider of Fig. 20-1 produces which of the following base voltages?   ( )
   (a) 0 V;   (b) −1.1 V;   (c) −1.8 V;   (d) −6.03 V.
2. The measured emitter voltage of Fig. 20-1 was closest to: ( )
   (a) 0 V;   (b) −1.1 V;   (c) −1.8 V;   (d) −6.03 V.
3. The measured collector voltage of Fig. 20-2 was closest to: ( )
   (a) 0 V;   (b) 3.97 V;   (c) 8.2 V;   (d) 8.9 V.
4. The emitter voltage measured in Fig. 20-2 was closest to: ( )
   (a) 0 V;   (b) slightly positive;   (c) slightly negative;   (d) 8.9 V.
5. With upside-down *pnp* bias, the emitter voltage is approximately: ( )
   (a) 0.7 V less than V_B;   (b) 0.7 V more than V_B;   (c) unknown;   (d) less than V_C.
6. What is the direction of current through each component of Fig. 20-2? The answers should be up, down, left, or right for each resistor and transistor.

TROUBLESHOOTING

7. Name all the troubles found that produced a collector voltage of 0 V.

8. In Fig. 20-2, suppose $V_B = 10$ V, $V_E = 10$ V, and $V_C = 0$. What is the trouble?

CRITICAL THINKING

9. Explain how the value of R_E was determined for the design.

10. Optional. Instructor's question.

Transistor Bias

Before a transistor can be used to amplify an ac signal, the quiescent (*Q*) point of operation is typically chosen to be near the middle of the dc load line. Then, the incoming ac signal can produce fluctuations above and below this *Q* point. The three most primitive forms of bias are base bias, emitter-feedback bias, and collector-feedback bias. These biasing methods are not the best ways to bias a transistor if a stable *Q* point is desired. Nevertheless, these biasing methods are occasionally used with small-signal amplifiers. In this experiment, all three types of biasing configurations will be built and analyzed to verify the operation as discussed in the textbook.

The most common transistor troubles are the collector-emitter short and collecter-emitter open. The collector-emitter short will be simulated by putting a jumper between the collector, base, and emitter; this is equivalent to shorting both diodes. The collector-emitter open will be simulated by removing the transistor from the circuit; this is equivalent to opening both diodes.

GOOD TO KNOW

The pinout for the 2N3904 is shown to the right. The actual data sheet can be found on the Internet. Search for the 2N3904 data sheet.

Emitter Base Collector

Required Reading

Chapter 7 (Sec. 7-9) of *Electronic Principles,* 8th ed.

Equipment

1   power supply: 15 V
3   transistors: 2N3904 (or equivalent)
7   ½-W resistors: 100 Ω, 680 Ω, 820 Ω, 1 kΩ, 220 kΩ, 270 kΩ, 470 kΩ
1   DMM (digital multimeter)

Procedure

BASE BIAS

1. Measure and record the values of the resistors. Locate a data sheet on the Internet for a 2N3904. Notice that the dc current gain h_{FE} has a minimum value of 100

and a maximum value of 300 for an I_C of 10 mA. The typical value is not listed. For this experiment, a typical value of 200 will be used.

2. In Fig. 21-1, use the typical h_{FE} to calculate I_B, I_C, and V_C. Record the answers in Table 21-1.

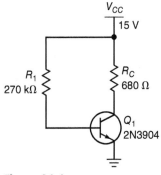

Figure 21-1

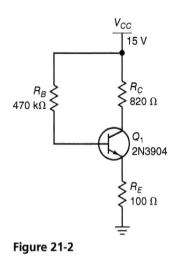

Figure 21-2

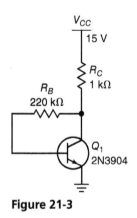

Figure 21-3

3. Build the circuit of Fig. 21-1. Measure and record the quantities listed in Table 21-1.
4. Repeat Steps 2 and 3 for the other transistors. If Multisim is being used, change the β_{dc} of transistors 2 and 3 to 250 and 150, respectively.

EMITTER-FEEDBACK BIAS

5. In Fig. 21-2, use the typical h_{FE} to calculate I_C, V_C, and V_E. Record the answers in Table 21-2.
6. Build the emitter-feedback bias of Fig. 21-2. Measure and record the quantities of Table 21-2.
7. Repeat Steps 5 and 6 for the other transistors.

COLLECTOR-FEEDBACK BIAS

8. In Fig. 21-3, use the typical h_{FE} to calculate and record the quantities of Table 21-3.
9. Build the circuit of Fig. 21-3. Measure and record all quantities listed in Table 21-3.
10. Repeat Steps 8 and 9 for the other transistors.

TROUBLESHOOTING

11. In Fig. 21-3, assume that the base resistor is open. Estimate and record the collector voltage V_C in Table 21-4.
12. Repeat Step 11 for the other troubles listed in Table 21-4.
13. Build the circuit of Fig. 21-3 with each trouble listed in Table 21-4. Measure and record the collector voltage.

CRITICAL THINKING

14. Design a collector-feedback biased circuit with a 2N3904 to meet the following specifications: $V_{CC} = 10$ V and $I_C = 2$ mA. *Hint:* Look at the 2N3904 data sheet. Calculate and record the quantities listed in Table 21-5.
15. Build the design. Measure and record the quantities of Table 21-5.

ADDITIONAL WORK (OPTIONAL)

16. Build the circuit of Fig. 21-1 using a 2N3906 and a collector supply voltage of -15 V. Measure V_B, V_E, and V_C. Record the value on a separate piece of paper.
17. Build the circuit of Fig. 21-2 using a 2N3906 and a collector supply voltage of -15 V. Measure and record V_B, V_E, and V_C.
18. Build the circuit of Fig. 21-3 using a 2N3906 and a collector supply voltage of -15 V. Measure and record V_B, V_E, and V_C.
19. Compare the recorded values of Steps 16 to 18 with the values recorded in Tables 21-1 to 21-3. What was learned about *npn* and *pnp* biasing circuits?

Experiment 21

Lab Partner(s) _____

PARTS USED Nominal Value	Measured Value	PARTS USED Nominal Value	Measured Value
100 Ω		220 kΩ	
680 Ω		270 kΩ	
820 Ω		470 kΩ	
1 kΩ			

CALCULATIONS

TABLE 21-1. BASE BIAS

	Calculated			Measured					
				Multisim			Actual		
Transistor	I_B	I_C	V_C	I_B	I_C	V_C	V_B	I_C	V_C
1									
2									
3									

TABLE 21-2. EMITTER-FEEDBACK BIAS

	Calculated			Measured					
				Multisim			Actual		
Transistor	I_C	V_C	V_E	I_C	V_C	V_E	I_C	V_C	V_E
1									
2									
3									

TABLE 21-3. COLLECTOR-FEEDBACK BIAS

	Calculated			Measured					
				Multisim			Actual		
Transistor	V_B	I_C	V_C	V_B	I_C	V_C	V_B	I_C	V_C
1									
2									
3									

TABLE 21-4. TROUBLESHOOTING

Trouble	Estimated V_C	Measured V_C	
		Multisim	Actual
Open 220 kΩ			
Shorted 220 kΩ			
Open 1 kΩ			
Shorted 1 kΩ			
Open C-E			
Shorted C-E			

TABLE 21-5. CRITICAL THINKING

Values: $R_B =$ _____; $R_C =$ _____

Transistor	Calculated V_C	Measured V_C			
		Multisim	% Error	Actual	% Error
1					
2					
3					

Questions for Experiment 21

1. Base bias has an unstable Q point because of the variation in:　　　　( )
   (a) base current;　　**(b)** V_{BE};　　**(c)** base resistance;　　**(d)** h_{FE}.
2. When the collector current increases in a base-biased circuit, the collector voltage:　( )
   (a) increases;　　**(b)** stays the same;　　**(c)** decreases.
3. In Fig. 21-2, the collector saturation current has a value of approximately:　( )
   (a) 5 mA;　　**(b)** 10 mA;　　**(c)** 15 mA;　　**(d)** 20 mA.
4. The measured data of Table 21-3 show that the V_{BE} drop was closest to:　( )
   (a) 0;　　**(b)** 0.3 V;　　**(c)** 0.7 V;　　**(d)** 7.85 V.
5. Of the three circuits tested, which had the most stable Q point?　( )
   (a) base bias;　　**(b)** emitter-feedback bias;　　**(c)** collector-feedback bias;
   (d) voltage-divider bias.
6. Briefly discuss the Q point for the three circuits tested.

TROUBLESHOOTING

7. While troubleshooting a circuit like Fig. 21-3, a collector voltage V_C of 15 V is measured. What are three possible troubles?

8. Name two possible troubles in Fig. 21-3 that would produce a collector voltage of zero.

CRITICAL THINKING

9. Explain how design values in Table 21-5 were calculated.

10. Optional. Instructor's question.

22

Coupling and Bypass Capacitors

Capacitive reactance decreases as frequency increases. Because of this, a capacitor has a large impedance at low frequencies and a small impedance at high frequencies. As an approximation, a capacitor is treated as a short for ac and an open for dc. When used in amplifiers, capacitors can couple the signal from one active node to another, or they can bypass the signal from an active node to ground.

The cutoff frequency is the frequency where $X_c = R$. From this, the following equation for the cutoff frequency can be derived:

$$f_c = \frac{1}{2\pi RC}$$

In Fig. 22-1a, $R = 68\ \text{k}\Omega \parallel 100\ \text{k}\Omega$. In Fig. 22-1b, $R = 22\ \text{k}\Omega \parallel 22\ \text{k}\Omega$. At the cutoff frequency, the output voltage is 0.707 times the input voltage.

GOOD TO KNOW

The capacitor and resistors form a voltage divider. The value X_C varies based on the frequency of V_{in} while the value of the resistors remains constant.

Required Reading

Chapter 8 (Secs. 8-1 and 8-2) of *Electronic Principles*, 8th ed.

Equipment

1 signal generator
4 ½-W resistors: two 22 kΩ, 68 kΩ, 100 kΩ
1 capacitor: 0.022 μF
1 oscilloscope

Procedure

1. Measure and record the capacitor and resistor values. Calculate the cutoff frequency in Fig. 22-1a. Fill in the values of f_c, $10f_c$, and $0.1f_c$ in Table 22-1 under f.
2. Build the circuit of Fig. 22-1a.
3. Adjust the signal generator to get a frequency of f_c and an input voltage v_{in} of 1 V peak-to-peak on the oscilloscope.

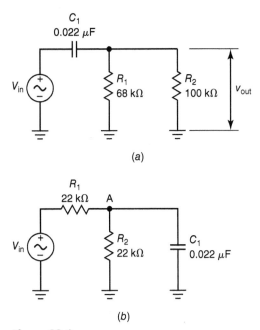

(a)

(b)

Figure 22-1

4. Measure the output voltage v_{out} and record in Table 22-1.
5. Change the frequency to $10f_c$ and readjust to get a v_{in} of 1 V_{p-p}. Measure v_{out} and record in Table 22-1.
6. Change the frequency to $0.1f_c$ and check that v_{in} is 1 V_{p-p}. Measure and record v_{out}.
7. Calculate the cutoff frequency in Fig. 22-1b. Fill in the values of f_c, $10f_c$, and $0.1f_c$ in Table 22-2.
8. Build the circuit of Fig. 22-1b without the capacitor. Adjust the frequency to f_c.
9. Set the signal level to 1 V_{p-p} across the lower 22 kΩ resistor.
10. Connect the capacitor between point A and ground. Then measure and record v_A.
11. Remove the capacitor and change the frequency to $10f_c$. Then repeat Steps 9 and 10.
12. Remove the capacitor and change the frequency to $0.1f_c$. Then repeat Steps 9 and 10.

CRITICAL THINKING (OPTIONAL)

13. This may require reading ahead in the textbook if Bode plots have not been covered yet. Construct an idealized Bode plot of the data in Tables 22-1 and 22-2 in a spreadsheet or on two-cycle, semilogarithmic graph paper. Information on Bode plots can be found in Chap. 14 of *Electronic Principles*. Use either two different colors or two different patterns for the lines on one set of axes. Attach the Bode plot to the report for this experiment.
14. Multisim has a virtual Bode plotter. Open the two Multisim experiment files labeled Critically Thinking Fig. 1a and 1b. Activate the simulation and observe the Bode plot generated for each circuit. Print both plots and attach them to the report for this experiment.

ADDITIONAL WORK (OPTIONAL)

15. Build the circuit of Fig. 22-1a with a small 8 Ω speaker connected across the 100 kΩ. Replace the 0.022 μF with 0.47 μF. Increase the output of the signal generator until an audio tone can be heard. Vary the frequency and notice how the tone changes pitch. Leave the output level of the signal generator fixed during the next step.
16. Build the circuit of Fig. 22-1b with a small 8 Ω speaker across the 0.022 μF. With the output level of the signal generator set the same as in the preceding step, what is heard?
17. What conclusions can be drawn from Steps 15 and 16? Why does this happen?

Experiment 22

Lab Partner(s) _____

PARTS USED Nominal Value	Measured Value	PARTS USED Nominal Value	Measured Value
22 kΩ		100 kΩ	
22 kΩ		0.022 μF	
68 kΩ			

CALCULATIONS

TABLE 22-1. COUPLING CAPACITOR

	f	Measured v_{out}	
		Multisim	Actual
f_c			
$10 f_c$			
$0.1 f_c$			

TABLE 22-2. BYPASS CAPACITOR

	f	Measured v_A	
		Multisim	Actual
f_c			
$10 f_c$			
$0.1 f_c$			

Questions for Experiment 22

1. A coupling capacitor ideally looks like a dc: ( )
   (a) open and ac open;   (b) open and ac short;   (c) short and ac open;
   (d) short and ac short.
2. A small Thevenin resistance means the bypass capacitor must be: ( )
   (a) small;   (b) large;   (c) unaffected;   (d) open.
3. The value of f_c in Table 22-1 is closest to: ( )
   (a) 100 Hz;   (b) 180 Hz;   (c) 1 kHz;   (d) 5 kHz.
4. The value of f_c in Table 22-2 is closest to: ( )
   (a) 100 Hz;   (b) 500 Hz;   (c) 650 Hz;   (d) 1 kHz.
5. If an f_c of 18 Hz in Fig. 22-1a is needed, the capacitor value must be changed to ( )
   approximately:
   (a) 0.1 μF;   (b) 0.2 μF;   (c) 0.5 μF;   (d) 1 μF.
6. In Table 22-1, the output voltage at f_c is closest to: ( )
   (a) 0.707 V;   (b) 0.9 V;   (c) 0.99 V;   (d) 1 V.
7. In Table 22-2, the output voltage at f_c is closest to: ( )
   (a) 0.01 V;   (b) 0.15 V;   (c) 0.707 V;   (d) 1 V.
8. When the input voltage is 1 V in Fig. 22-1a, the output voltage at $10f_c$ is closest to: ( )
   (a) 0;   (b) 0.707 V;   (c) 0.9 V;   (d) 1 V.

9. Optional. Instructor's question.

10. Optional. Instructor's question.

The CE Amplifier

After the transistor of a CE amplifier has been biased with its Q point near the middle of the dc load line, a small ac signal can be coupled into the base. This produces an amplified ac signal at the collector. In this experiment, a CE amplifier is built and the voltage gain measured, as well as looking at the dc and ac waveforms throughout the circuit.

GOOD TO KNOW

When dc coupled, the oscilloscope will display the ac waveform and any dc level the waveform is riding on. When ac coupled, only the ac waveform is displayed, with the dc level removed.

Required Reading

Chapter 8 (Sec. 8-9) of *Electronic Principles,* 8th ed.

Equipment

1   signal generator
1   power supply: 10 V
3   transistors: 2N3904 (or equivalent)
4   ½-W resistors: 1 kΩ, 2.2 kΩ, 3.9 kΩ, 10 kΩ
2   capacitors: 1 μF, 470 μF (10-V rating or higher)
1   oscilloscope

Procedure

DC AND AC VOLTAGES

1. Measure and record the capacitor and resistor values. In Fig. 23-1, calculate the dc voltage at the base, emitter, and collector. Record the answers in Table 23-1.
2. Calculate and record the peak-to-peak ac voltages at the base, emitter, and collector.
3. Connect the circuit. Adjust the signal generator to get an input signal of 10 mV peak-to-peak at 1 kHz. If the function generator doesn't produce a clean waveform at that amplitude, build a voltage divider to reduce the amplitude.

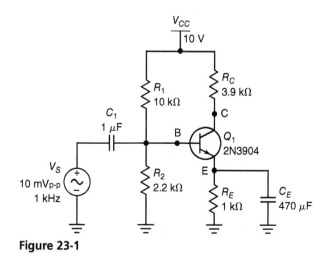

Figure 23-1

4. Use the oscilloscope on dc coupled to measure dc voltage and on ac coupled to measure the ac peak-to-peak voltage at the base, emitter, and collector of the transistor. Record all voltages in Table 23-1.
5. With the oscilloscope on dc input, waveforms like those of Fig. 8-8 in the textbook should be seen. This confirms that the total voltages are the sum of dc and ac components.

PHASE INVERSION

6. Using a dual-trace oscilloscope, look at the base signal with one input and the collector signal with the other input. Also, use the collector signal to drive the trigger of the oscilloscope. Notice that the collector signal is 180° out of phase with the base signal.
7. If a single-trace oscilloscope is being used, externally trigger the oscilloscope with the collector signal. (If in doubt, ask the instructor about this.) Look first at the base signal, then at the collector signal. Notice that the signals are 180° out of phase.

VOLTAGE GAIN

8. In Fig. 23-1, use Eq. (8-10) in the textbook to calculate the ideal emitter resistance r'_e. Use Eq. (8-16) to calculate the voltage gain A_V. Record the answers in Table 23-2.
9. Build the circuit with any of the three transistors. Measure and record the input and output ac voltages.
10. Calculate the actual voltage gain using the v_{out} and v_{in} measured in Step 9. Next calculate r'_e using the ratio R_C/A_V. Record the experimental A_V and r'_e in Table 23-2.
11. Repeat Steps 8 to 10 for the other transistors. If Multisim is being used, change the β_{dc} of transistors 2 and 3 to 250 and 150, respectively.

TROUBLESHOOTING

12. In Fig. 23-1, assume that C_1 is open. Estimate the peak-to-peak ac voltage at the base, emitter, and collector. Record in Table 23-3.
13. Repeat Step 12 for each trouble listed in Table 23-3.
14. Build the circuit with each trouble. Measure and record the ac voltages.

CRITICAL THINKING

15. Select a value of collector resistance in Fig. 23-1 to produce a theoretical voltage gain of 100. Using the nearest standard resistance, calculate and record the quantities of Table 23-4.
16. Build the circuit with the design value of R_C. Measure and record the quantities listed in Table 23-4.
17. Repeat Step 16 for the other transistors.

ADDITIONAL WORK (OPTIONAL)

18. Have another student insert one of the following troubles into the circuit: open or short any resistor, open any capacitor, open any connecting wire, short the BE, CB, or CE terminals of the transistor. Use only voltage readings of a DMM or an oscilloscope to troubleshoot.
19. Repeat Step 18 several times to build confidence in being able to troubleshoot the circuit for various troubles.

Experiment 23

Lab Partner(s) _____

PARTS USED Nominal Value	Measured Value	PARTS USED Nominal Value	Measured Value
1 kΩ		10 kΩ	
2.2 kΩ		1 μF	
3.9 kΩ		470 μF	

CALCULATIONS

TABLE 23-1. CE AMPLIFIER

	Calculated			Measured					
				Multisim			Actual		
	B	E	C	B	E	C	B	E	C
dc									
ac									

TABLE 23-2. VOLTAGE GAIN

	Calculated		Measured				Experimental	
			Multisim		Actual			
Transistor	r_e'	A_V	v_{in}	v_{out}	v_{in}	v_{out}	A_V	r_e'
1								
2								
3								

TABLE 23-3. TROUBLESHOOTING

	Estimated			Measured					
				Multisim			Actual		
Trouble	v_b	v_e	v_c	v_b	v_e	v_c	v_b	v_e	v_c
Open C_1									
Open R_2									
Open R_E									

TABLE 23-4. CRITICAL THINKING

	Calculated			Measured							
				Multisim				Actual			
Transistor	r_e'	R_C	A_V	v_{in}	v_{out}	A_V	% Error	v_{in}	v_{out}	A_V	% Error
1											
2											
3											

Questions for Experiment 23

1. The CE amplifier of Fig. 23-1 has a theoretical r'_e of: (   )
   (a) 22.7 Ω;   (b) 1 kΩ;   (c) 3.6 kΩ;   (d) 10 kΩ.

2. Ideally, the CE amplifier of Fig. 23-1 has a voltage gain of approximately: (   )
   (a) 1;   (b) 3.6;   (c) 4.54;   (d) 159.

3. The emitter of Fig. 23-1 had little or no ac signal because of: (   )
   (a) the emitter resistor;   (b) the input coupling capacitor;   (c) the emitter bypass capacitor;   (d) the weak base signal.

4. The voltage at collector was closest to: (   )
   (a) 6 V dc and 10 mV ac;   (b) 1.8 V dc and 1.6 V ac;   (c) 1.1 V dc and 10 mV ac;   (d) 6 V dc and 1.6 V ac.

5. The dc bias of the transistor is undisturbed by the dc resistance of the signal generator because the input coupling capacitor: (   )
   (a) blocks dc;   (b) transmits ac;   (c) blocks ac;   (d) transmits dc.

6. As briefly as possible, explain how the circuit of this experiment amplifies the signal.

TROUBLESHOOTING

7. What happens to the dc voltages of the circuit when the coupling capacitor is open? What happens to the ac voltages?

8. Explain the measured voltages when R_E is open.

CRITICAL THINKING

9. Explain how the value of collector resistance is selected.

10. Optional. Instructor's question.

Other CE Amplifiers

Because of the input impedance of a CE amplifier, some of the ac signal may be dropped across the source impedance. Furthermore, the Thevenin equivalent of the amplifier output is an ac generator in series with the output impedance of the amplifier. When a load resistance is connected to the amplifier, some of the ac signal is dropped across the output impedance.

One way to increase the input impedance is to use a swamping resistor in the emitter circuit. This also stabilizes the voltage gain against changes in r'_e. Because the swamping resistor reduces the voltage gain, it may be necessary to cascade two swamped amplifiers to get the same voltage gain as a single unswamped stage.

GOOD TO KNOW

Signal generators often produce a noisy signal at very small amplitudes. A 100:1 voltage divider can be used when a small input signal is required.

Required Reading

Chapter 8 (Secs. 8-9 to 8-11) of *Electronic Principles,* 8th ed.

Equipment

1 signal generator
1 power supply: 10 V
3 transistors: 2N3904 (or equivalent)
9 ½-W resistors: 180 Ω, 820 Ω, two 1 kΩ, 1.5 kΩ, 2.2 kΩ, 3.9 kΩ, 10 kΩ, 51 kΩ
3 capacitors: two 1 μF, 470 μF (10-V rating or higher)
1 oscilloscope

Procedure

CE AMPLIFIER WITH SOURCE AND LOAD RESISTANCES

1. Measure and record the capacitor and resistor values. In Fig. 24-1, assume h_{fe} (same as β) is 150. Calculate the input impedance of the stage. Also calculate the peak-to-peak base voltage and the peak-to-peak collector voltage. Record the answers in Table 24-1.

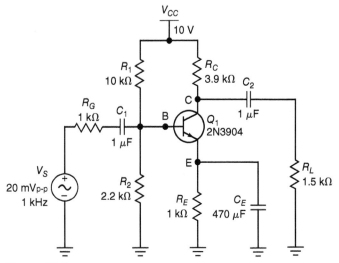

Figure 24-1

2. Build the circuit. Adjust the signal generator to get a source signal of 20 mV peak-to-peak at 1 kHz. (Measure this between the left end of the source resistance and ground.)

3. Use the oscilloscope on dc coupling to view the waveforms at the base and collector, and verify that the waveforms are the sum of dc and ac components. Also view the waveform at the emitter. Because of the bypass capacitor, the emitter should have only a dc component.
4. With the oscilloscope on ac coupling, measure the peak-to-peak voltage at the base and collector. Record the data in Table 24-1.
5. Repeat Step 4 for the two other transistors. If Multisim is being used, change the β of transistors 2 and 3 to 200 and 100, respectively.

SWAMPED AMPLIFIER

6. In Fig. 24-2, assume an h_{fe} of 150 and calculate the input impedance of the stage. Also calculate the peak-to-peak voltage at the base and collector. Record the answers in Table 24-2.
7. Build the circuit. Measure and record the peak-to-peak voltage at the base and collector.
8. Repeat Steps 6 and 7 for the other two transistors.

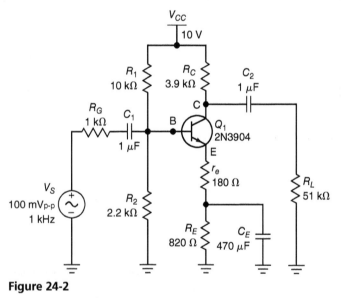

Figure 24-2

TROUBLESHOOTING

9. In Fig. 24-2, assume that C_E is open. Estimate the peak-to-peak ac voltage at the base, emitter, and collector. Record in Table 24-3.
10. Repeat Step 9 for each trouble listed in Table 24-3.
11. Build the circuit with each trouble. Measure and record the ac voltages.

CRITICAL THINKING

12. Select a value of swamping resistance in Fig. 24-2 to produce an unloaded voltage gain of 10 from the base to the collector. Using the nearest standard resistance, calculate and record the quantities of Table 24-4.
13. Build the circuit with the selected value of r_E. Measure and record the quantities listed in Table 24-4.
14. Repeat Step 13 for the other transistors.

ADDITIONAL WORK (OPTIONAL)

15. Have another student insert one of the following troubles into the circuit of Fig. 24-2: open or short any resistor; open any capacitor; open any connecting wire; short the BE, CB, or CE terminals of the transistor. Use only voltage readings of a DMM or an oscilloscope to troubleshoot.
16. Repeat Step 15 several times to build confidence in being able to troubleshoot the circuit for various troubles.

Experiment 24

Lab Partner(s) _____

PARTS USED Nominal Value	Measured Value	PARTS USED Nominal Value	Measured Value	PARTS USED Nominal Value	Measured Value
180 Ω		1.5 kΩ		51 kΩ	
820 Ω		2.2 kΩ		1 μF	
1 kΩ		3.9 kΩ		1 μF	
1 kΩ		10 kΩ		470 μF	

TABLE 24-1. CE AMPLIFIER WITH SOURCE AND LOAD RESISTANCES

	Calculated			Measured				
				Multisim		Actual		
Transistor	z_{in}	v_b	v_c	v_b	v_c	v_b	v_c	
1								
2								
3								

TABLE 24-2. SWAMPED AMPLIFIER

	Calculated			Measured				
				Multisim		Actual		
Transistor	z_{in}	v_b	v_c	v_b	v_c	v_b	v_c	
1								
2								
3								

TABLE 24-3. TROUBLESHOOTING

	Estimated			Measured							
				Multisim			Actual				
Trouble	v_b	v_e	v_c	v_b	v_e	v_c	v_b	v_e	v_c		
Open C_E											
Shorted C_E											
Open C-E											
Shorted C-E											
Open C_2											
Shorted C_2											

TABLE 24-4. CRITICAL THINKING

| Transistor | Calculated | | | Measured | | | | | | |
| | r_E | v_b | v_c | Multisim | | | Actual | | | |
				v_b	v_c	% Error	v_b	v_c	% Error	
1										
2										
3										

Questions for Experiment 24

1. The calculated base voltage in Table 24-1 is approximately: ( )
   (a) 10.8 mV;   (b) 20 mV;   (c) 250 mV;   (d) 500 mV.
2. The calculated collector voltage in Table 24-1 was closest to: ( )
   (a) 10.8 mV;   (b) 20 mV;   (c) 250 mV;   (d) 500 mV.
3. In Table 24-2, the measured base voltage was closest to: ( )
   (a) 12 mV;   (b) 20 mV;   (c) 63 mV;   (d) 1 V.
4. The voltage gain from base to collector in the swamped amplifier was closest to: ( )
   (a) 1;   (b) 5;   (c) 10;   (d) 15.
5. Compared with the CE amplifier, the swamped amplifier had a: ( )
   (a) lower input impedance;   (b) higher output impedance;   (c) lower
   voltage gain;   (d) lower ac collector voltage.
6. Explain why an amplifier with a swamping resistor has a more stable voltage gain than one
   without.

TROUBLESHOOTING

7. What happens to the voltage gain of an amplifier when the emitter bypass capacitor is open?
   Explain why this happens.

8. Explain what happens when the emitter bypass capacitor is shorted.

CRITICAL THINKING

9. How was the value of the swamping resistor determined?

10. Optional. Instructor's question.

Cascaded CE Stages

The amplified signal from a CE stage can be used as the input to another CE stage. In this way, a multistage amplifier with very large voltage gain can be built. Because a CE stage has an input impedance, there is a loading effect on the preceding stage. In other words, the loaded voltage gain is less than the unloaded voltage gain. In this experiment, a two-stage amplifier using swamped stages to stabilize the overall voltage gain will be built and analyzed.

GOOD TO KNOW

Signal generators often produce a noisy signal at very small amplitudes. A 100:1 voltage divider can be used when a small input signal is required.

Required Reading

Chapter 9 (Secs. 9-1 and 9-2) of *Electronic Principles,* 8th ed.

Equipment

- 1 signal generator
- 1 power supply: 10 V
- 2 transistors: 2N3904 (or equivalent)
- 12 ½-W resistors: two 68 Ω, three 1 kΩ, one 1.2 kΩ, two 2.2 kΩ, two 3.9 kΩ, two 10 kΩ
- 5 capacitors: three 1 μF, two 47 μF (10-V rating or higher)
- 1 DMM (digital multimeter)
- 1 oscilloscope

Procedure

CALCULATIONS

1. Measure and record the resistor and capacitor values. In Fig. 25-1, calculate the dc voltages at the base, emitter, and collector of each stage. Record the answers in Table 25-1.

2. Next, locate a data sheet for a 2N3904 on the Internet. Read the typical value of h_{fe}. Record this value at the top of Table 25-2.

3. Calculate the peak-to-peak ac voltage at the base, emitter, and collector of each stage (Fig. 25-1) using the h_{fe} of Step 2. Record all ac voltages in Table 25-2.

TESTS

4. Build the two-stage amplifier of Fig. 25-1.

5. Measure the dc voltage at the base, emitter, and collector of each stage. Record the data in Table 25-1. Within the tolerance of the resistors being used, the measured voltages should agree with the calculated voltages.

6. Measure the peak-to-peak ac voltage at the base, emitter, and collector of each stage. Record the data in Table 25-2. These measured ac voltages should agree with the calculated values.

LOADING EFFECTS

7. Open the coupling capacitor between the first and second stage. Use the oscilloscope to view the amplitude of the ac waveform on the first collector. Reconnect the coupling capacitor and notice that the ac signal decreases significantly. By removing the coupling

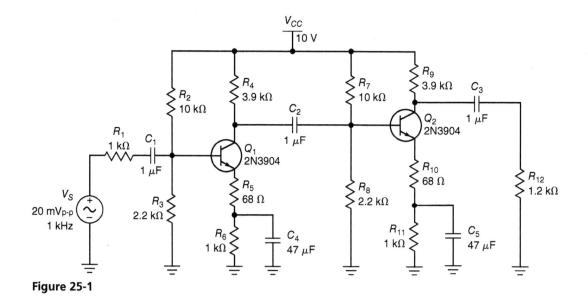

Figure 25-1

capacitor, the loading effect of the second stage is removed from the output of the first stage.

8. Open the coupling capacitor between the second stage and the load resistor. Use the oscilloscope to view the amplitude of the ac waveform on the second collector. Reconnect the coupling capacitor and notice the decrease in signal amplitude. The effect of the load resistor on the second stage is comparable to what was observed in Step 7.

TROUBLESHOOTING

9. In Fig. 25-1, assume that C_4 is open. Does this produce a trouble in the first or second stage? Record the answer (1 or 2) in Table 25-3.
10. Insert the foregoing trouble in the circuit. Use a DMM to measure the dc voltages and an oscilloscope to measure the ac voltages. Before each measurement, make a ballpark estimate of its value. Then when the voltage is measured, it can be compared against the estimated value to determine if a fault is present.
11. Estimate each voltage in Table 25-3 for the stage with the trouble. Measure and record the voltage.
12. Repeat Steps 9 to 11 for each of the troubles listed in Table 25-3.

CRITICAL THINKING

13. Select a swamping resistor for the second stage to get an overall voltage gain of approximately 75. Record the value at the top of Table 25-4.
14. Change the swamping resistor to the design value. Measure and record the voltage gain of the first stage (base to collector). Then measure and record the voltage gain of the second stage.
15. Measure and record the overall voltage gain (first base to second collector).

APPLICATION (OPTIONAL)

16. Build the circuit shown in Fig. 25-2. The +5 V supply V_{DD}, R_1, and the switch S_1 simulate a TTL output.
17. Measure the dc input voltage of Q_1 and the output dc voltage of Q_2 for the two switch positions. The dc output voltage of Q_2 should match V_{CC} when the input dc voltage of Q_1 is +5 V.
18. The current provided to the load can vary, based on the value of R_3.

ADDITIONAL WORK (OPTIONAL)

19. Have another student insert one of the following troubles into the circuit of Fig. 25-1: open or short any resistor; open any capacitor; open any connecting wire; short the BE, CB, or CE terminals of either transistor. Use only voltage readings of a DMM or an oscilloscope to troubleshoot.
20. Repeat Step 19 several times to build confidence in being able to troubleshoot the circuit for various troubles.

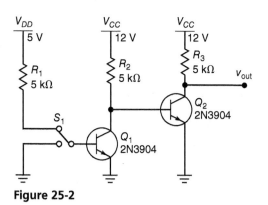

Figure 25-2

Experiment 25

Lab Partner(s) _____

PARTS USED Nominal Value	Measured Value	PARTS USED Nominal Value	Measured Value	PARTS USED Nominal Value	Measured Value
68 Ω		2.2 kΩ		1 μF	
68 Ω		2.2 kΩ		1 μF	
1 kΩ		3.9 kΩ		1 μF	
1 kΩ		3.9 kΩ		47 μF	
1 kΩ		10 kΩ		47 μF	
1.2 kΩ		10 kΩ			

TABLE 25-1. DC VOLTAGES

	Calculated			Measured					
				Multisim			Actual		
Stage	V_B	V_E	V_C	V_B	V_E	V_C	V_B	V_E	V_C
1									
2									

TABLE 25-2. AC VOLTAGES: h_{fe} = _____

	Calculated			Measured					
				Multisim			Actual		
Stage	v_b	v_e	v_c	v_b	v_e	v_c	v_b	v_e	v_c
1									
2									

TABLE 25-3. TROUBLESHOOTING

		Measured DC Voltages						Measured AC Voltages					
		Multisim			Actual			Multisim			Actual		
Trouble	Stage	V_B	V_E	V_C	V_B	V_E	V_C	v_b	v_e	v_c	v_b	v_e	v_c
Open C_4													
Shorted C_4													
Shorted R_{10}													
Open R_3													
Open C_5													

TABLE 25-4. CRITICAL THINKING: $r_E =$ _____

	Multisim	% Error	Actual	% Error
A_{v1}				
A_{v2}				
A_{v3}				

Questions for Experiment 25

1. The calculated dc base voltage of the first stage was approximately:　　　　( )
   (a) 1.1 V;　　(b) 1.8 V;　　(c) 6.28 V;　　(d) 10 V.
2. The measured dc collector voltage of the second stage was closest to:　　( )
   (a) 1.1 V;　　(b) 1.8 V;　　(c) 6.28 V;　　(d) 10 V.
3. The ac base voltage of the first stage was closest to:　　　　　( )
   (a) 5 mV;　　(b) 12 mV;　　(c) 100 mV;　　(d) 1.4 V.
4. The ac emitter voltage of the second stage was closest to:　　　( )
   (a) 5 mV;　　(b) 12 mV;　　(c) 100 mV;　　(d) 1.4 V.
5. The voltage gain from the base of the first stage to the collector of the second stage　( )
   was closest to:
   (a) 10;　　(b) 115;　　(c) 230;　　(d) 1000.
6. Explain why the signal decreased when the coupling capacitor was reconnected in Step 7.

TROUBLESHOOTING

7. Explain what happens when an emitter bypass capacitor opens.

8. Suppose the collector-emitter is shorted in the first stage of Fig. 25-1. What is the approximate input impedance looking into the base of the first stage? Explain the answer.

CRITICAL THINKING

9. There is a simple way to modify Fig. 25-1 for use with *pnp* transistors. Explain what changes need to be made.

10. Optional. Instructor's question.

CC and CB Amplifiers

An emitter follower or common-collector (CC) amplifier has a high input impedance, low output impedance, and low nonlinear distortion. It is often used as a buffer stage between a high-impedance source and a low-resistance load. A common-base (CB) amplifier has a low input impedance and relatively high output impedance, and is used as a high-frequency voltage amplifier. In this experiment, CC and CB amplifier circuits will be built and their circuit characteristics examined.

GOOD TO KNOW

The typical range of h_{fe} for a 2N3904 transistor is 100 to 300. Another name for h_{fe} is β_{dc}.

Required Reading

Chapter 9 (Secs. 9-3 to 9-8) of *Electronic Principles*, 8th ed.

Equipment

1 signal generator
1 power supply: 10 V
1 transistor: 2N3904 (or equivalent)
8 ½-W resistors: 51 Ω, two 2 kΩ, 3.6 kΩ, 3.9 kΩ, 4.7 kΩ, two 10 kΩ
4 capacitors: 1 μF, two 10 μF, 470 μF (10-V rating or higher)
1 DMM (digital multimeter)
1 oscilloscope

Procedure

EMITTER FOLLOWER (CC)

1. Measure and record the resistor and capacitor values. In Fig. 26-1, calculate the dc voltage at the base, emitter, and collector. Record the answers in Table 26-1.

2. Locate a 2N3904 data sheet on the Internet to determine the typical h_{fe}. Record this value in Table 26-1.
3. Calculate and record the peak-to-peak ac voltage at the base, emitter, and collector.
4. Build the circuit. Measure and record the dc voltage at the base, emitter, and collector.

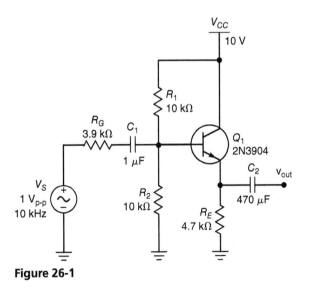

Figure 26-1

5. Adjust the signal generator to get a source signal of 1 V peak-to-peak at 10 kHz. (Measure this between the left end of the source resistor and ground.)
6. Measure and record the peak-to-peak ac voltage at the base, emitter, and collector.

CC OUTPUT IMPEDANCE

7. Calculate the theoretical output impedance for the circuit of Fig. 26-1. Record the answer in Table 26-2.
8. Reduce the peak-to-peak source from 1 V to 100 mV.
9. Measure and record the peak-to-peak output voltage (unloaded).
10. Connect a load resistance of 51 Ω across the output.
11. Measure the loaded output voltage.
12. Calculate the output impedance of the emitter follower with the data of Steps 9 and 11. Record the experimental answer in Table 26-2.

TROUBLESHOOTING

13. In Fig. 26-1, assume that R_1 is open. Estimate the dc and ac voltages at the output emitter. Record the answers in Table 26-3.
14. Repeat Step 13 for each trouble listed in Table 26-3.
15. Build the circuit with each trouble. Measure and record the dc and ac voltages.

COMMON-BASE (CB)

16. In Fig. 26-2, calculate the dc voltage of the base, emitter, and collector. Record the answers in Table 26-4.
17. Build the circuit. Measure and record the dc voltage at the base, emitter, and collector.
18. Adjust the signal generator to get a source signal of 20 mV peak-to-peak at point A.

19. Calculate and record the peak-to-peak ac signal at the base, emitter, and collector.
20. Measure and record the ac signal voltage at the base, emitter, and collector. Make note of the phase shift between the input (emitter) and the output (collector).

CB INPUT IMPEDANCE

21. Calculate the theoretical input impedance for the circuit of Fig. 26-2. Record the answer in Table 26-5.
22. Connect a 51 Ω resistor between the output of the signal generator and C_1.
23. Adjust the output of the signal generator to 20 mV peak-to-peak.
24. Measure the ac signal at the emitter.
25. Calculate the input impedance of the amplifier using the data of Steps 23 and 24. Record the experimental answer in Table 26-5.

CRITICAL THINKING

26. In Fig. 26-2, select a value of R_1 to double the circuit's voltage gain. Record the design value in Table 26-6.
27. Build the circuit with the design value of R_1. Measure and record the ac output voltage.

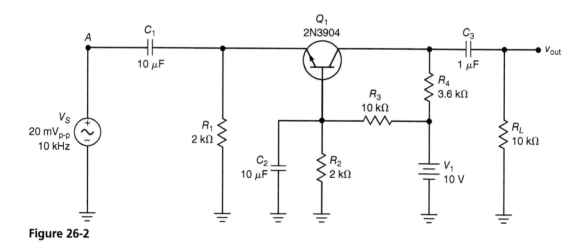

Figure 26-2

Experiment 26

Lab Partner(s) _____

PARTS USED Nominal Value	Measured Value	PARTS USED Nominal Value	Measured Value	PARTS USED Nominal Value	Measured Value
51 Ω		3.9 kΩ		1 μF	
2 kΩ		4.7 kΩ		10 μF	
2 kΩ		10 kΩ		10 μF	
3.6 kΩ		10 kΩ		470 μF	

TABLE 26-1. EMITTER FOLLOWER: h_{fe} = _____

	Calculated			Measured					
				Multisim			Actual		
	B	E	C	B	E	C	B	E	C
dc									
ac									

TABLE 26-2. CC OUTPUT IMPEDANCE CALCULATED z_{out} = _____

	Multisim	Actual
Unloaded v_{out}		
Loaded v_{out}		
Experimental z_{out}		

TABLE 26-3. TROUBLESHOOTING

	Estimated		Measured			
			Multisim		Actual	
Trouble	V_E	V_e	V_E	V_e	V_E	V_e
Open R_1						
Shorted R_1						
Open R_2						
Shorted R_2						
Open R_E						
Shorted R_E						

TABLE 26-4. COMMON-BASED (CB)

	Calculated			Measured					
				Multisim			Actual		
	B	E	C	B	E	C	B	E	C
dc									
ac									

TABLE 26-5. CB INPUT IMPEDANCE

	Experimental z_{in}	
Calculated z_{in}	Multisim	Actual

TABLE 26-6. CRITICAL THINKING R_1 = _____

	Calculated	Measured			
		Multisim	% Error	Actual	% Error
v_{out}					
A_v					

Questions for Experiment 26

1. The data of Table 26-1 show that the voltage gain of the emitter follower was approximately: ( )
   (a) 0;   (b) 1;   (c) 4.3 V;   (d) 10 V.
2. The ac collector voltage of the emitter follower was closest to: ( )
   (a) 0;   (b) 0.58 V;   (c) 1 V;   (d) 10 V.
3. Because the ac emitter voltage approximately equals the ac base voltage in Table 26-1, the input impedance of the base must be: ( )
   (a) 0;   (b) very low;   (c) 10 V;   (d) very high.
4. The calculated input impedance in Table 26-5 is closest to: ( )
   (a) 1 Ω;   (b) 23 Ω;   (c) 42.4 Ω;   (d) 50 Ω.
5. The voltage gain in Fig. 26-2 is closest to: ( )
   (a) 10;   (b) 15;   (c) 25;   (d) 50.
6. Explain how the experimental value of z_{in} in Table 26-5 was determined.

TROUBLESHOOTING

7. In Fig. 26-1, explain the dc and ac output voltage when R_2 was shorted.

8. If there is a collector-emitter short in Fig. 26-1, what happens to the input impedance of the emitter follower?

CRITICAL THINKING

9. In Fig. 26-2, how was the design value of R_1 determined?

10. Optional. Instructor's question.

Emitter Follower Applications

The emitter follower has a high input impedance, low output impedance, and a current gain approximately equal to β. Because of this, the emitter follower is commonly used as a buffer. By connecting two transistors in a Darlington configuration, the overall current gain becomes very high and can be used to control large currents.

By cascading a zener diode and an emitter follower, voltage regulators capable of handling large load currents can be constructed. An improved regulator like this can hold the load voltage almost constant despite large changes in load current.

In this experiment, an emitter-follower speaker driver, a Darlington connected dc motor driver, and a regulated power supply will be built and investigated. The transformer should have a fused line cord with all primary connections insulated to avoid electrical shock. Be sure to ask your instructor for help, if you are unsure of how to properly connect the transformer.

GOOD TO KNOW

In addition to the high z_{in} and low z_{out}, the output waveform of the emitter follower is in phase with the input waveform.

Required Reading

Chapter 9 (Secs. 9-3 to 9-8) of *Electronic Principles*, 8th ed.

Equipment

1 center-tapped transformer, 12.6 Vac (Triad F-44× or equivalent) with fused line cord
4 silicon diodes: 1N4001 (or equivalent)
1 zener diode: 1N757 (or equivalent 9-V zener diode)
2 transistors: 2N3904, 2N3055 (or equivalent power transistor)
6 ½-W resistors: 100 Ω, 220 Ω, 820 Ω, 1 kΩ, 2.2 kΩ
2 1-W resistors: 100 Ω, 820 Ω
1 potentiometer: 10 kΩ
2 capacitors: 1 μF and 470 μF (16 V or higher)

1 speaker: 8 Ω
1 dc motor: MCM Part #: 28-12810 (or equivalent dc motor)
1 DMM (digital multimeter)
1 oscilloscope

Procedure

EMITTER FOLLOWER SPEAKER DRIVER

1. Measure and record the resistor and capacitor values. In Fig. 27-1, calculate the dc voltage at the base, emitter, and collector. Record the answers in Table 27-1.
2. Calculate the peak-to-peak ac voltage across the speaker and the circuit's voltage gain. Record the answers in Table 27-1.

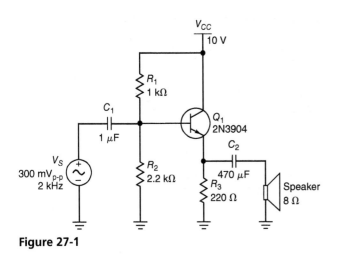

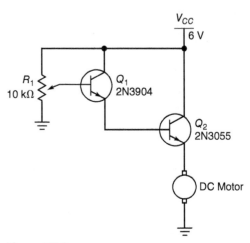

Figure 27-1

Figure 27-2

3. Build the circuit shown in Fig. 27-1 using a small 8-Ω speaker. Adjust the frequency of the signal generator to 2 kHz. Adjust the output level of the signal generator to produce an ac base voltage of 300 mV$_{p-p}$. A tone should be present from the speaker. Vary the frequency and notice that the tone changes.

4. Measure the dc transistor voltages and the ac voltage across the speaker. Using measured values, calculate the voltage gain of the emitter follower. Record these values in Table 27-1.

DARLINGTON MOTOR DRIVER

5. The MCM Part #: 28-12810 is a small dc motor that has the following specifications: 6 to 24 V dc input with a full-load current of 726 mA and a no-load current of 180 mA. The motor will be run under no-load conditions since it will not be connected to a mechanical load.

6. In Fig. 27-2, the 2N3055 is a power transistor that can produce enough output current for the motor.

7. Build the circuit of Fig. 27-2. When the dc power supply is turned on, use the potentiometer to set the base voltage of Q_1 to 3 V. The motor shaft should begin to spin.

8. With V_B set to 3 V, measure I_{B1}, I_{B2}, and I_{motor}. Record these values in Table 27-2.

9. Using the measured current values, calculate and record β_1, β_2, and overall β values in Table 27-2.

10. Firmly attach a piece of electrical tape to the motor shaft, leaving a ½-inch flap. Adjust the potentiometer to vary the voltage at the base of Q_1 from 1 V to 5 V. The motor should change speed from off to fully on. Now, run the motor at a very slow speed. The tape should indicate the direction of rotation.

11. Reverse the motor leads. The motor shaft will turn in the opposite direction. If the motor speed is decreased down to zero, the tape will be seen turning in the opposite direction.

REGULATED POWER SUPPLY

12. A 1N757 has a nominal zener voltage of 9.1 V. In Fig. 27-3, calculate the input voltage, zener voltage, and output voltage for the zener follower. (The input voltage is across the filter capacitor.) Record the answers in Table 27-3.

13. Build the regulated power supply of Fig. 27-3.

14. Measure and record all voltages listed in Table 27-3.

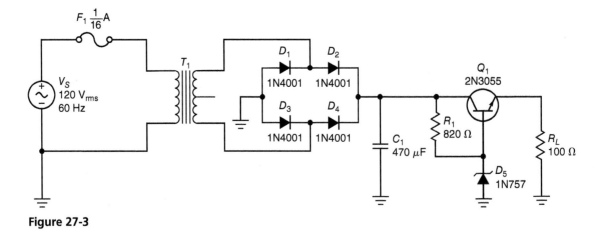

Figure 27-3

VOLTAGE REGULATION

15. Estimate and record the output voltages in Fig. 27-3 for each of the load resistors listed in Table 27-4.
16. Measure and record the output voltages for the load resistances of Table 27-4.

RIPPLE ATTENUATION

17. For each load resistance listed in Table 27-5, calculate and record the peak-to-peak ripple across the filter capacitor. Also calculate and record the peak-to-peak ripple at the output. (Assume a zener resistance of 10 Ω.)
18. For each load resistance of Table 27-5, measure and record the peak-to-peak ripple at the input and output of the zener follower. (*Note:* If the ripple appears very fuzzy or erratic, there may be oscillations. Try shortening the leads. If that doesn't work, consult the instructor.)

TROUBLESHOOTING

19. In Fig. 27-3, assume that D_1 is open.
20. Estimate the input and output voltage of the zener follower for the foregoing trouble. Record the answers in Table 27-6.
21. Build the circuit with the foregoing trouble. Measure and record the input and output voltage.
22. Repeat Steps 20 and 21 for the other troubles listed in Table 27-6.

ADDITIONAL WORK (OPTIONAL)

23. Have another student insert the following trouble into the circuit: open any component or connecting wire. Use only voltage readings of a DMM or an oscilloscope to troubleshoot.
24. Repeat Step 23 several times to build confidence in being able to troubleshoot the circuit for various troubles.
25. Measure the load voltage and load current for several values of load resistance between 100 Ω and 10 kΩ. Record the data on a separate piece of paper.
26. Using the data of Step 25, graph V_L versus I_L on semi-log paper.
27. Using the data of Step 25, graph V_L versus R_L on semi-log paper.

Experiment 27

Lab Partner(s) _____

PARTS USED Nominal Value	Measured Value	PARTS USED Nominal Value	Measured Value	PARTS USED Nominal Value	Measured Value
100 Ω		820 Ω		2.2 kΩ	
100 Ω		820 Ω		1 μF	
220 Ω		1 kΩ		470 μF	

TABLE 27-1. SPEAKER DRIVER

		Measured	
	Calculated	Multisim	Actual
V_B			
V_E			
V_C			
$v_{speaker}$			
A_V			

TABLE 27-2. DARLINGTON MOTOR DRIVER

	V_B	I_{B1}	I_{B2}	I_{motor}	β_1	β_2	β
Multisim							
Actual							

TABLE 27-3. REGULATED POWER SUPPLY

		Measured	
	Calculated	Multisim	Actual
V_{in}			
V_Z			
V_{out}			

TABLE 27-4. VOLTAGE REGULATION

		Measured V_{out}	
R_L	Estimated	Multisim	Actual
100 Ω			
1 kΩ			
10 kΩ			

TABLE 27-5. RIPPLE

R_L	Calculated V_{rip}		Measured V_{rip}			
			Multisim		Actual	
	In	Out	In	Out	In	Out
100 Ω						
1 kΩ						
10 kΩ						

TABLE 27-6. TROUBLESHOOTING

Trouble	Estimated		Measured			
			Multisim		Actual	
	V_{in}	V_{out}	V_{in}	V_{out}	V_{in}	V_{out}
Open D_1						
Open R_1						
Shorted D_5						
C-E open						

Questions for Experiment 27

1. When the load resistance increases in Table 27-4, the measured output voltage: ( )
   (a) decreases slightly;   (b) stays the same;   (c) increases slightly;
   (d) none of the foregoing. ( )
2. When the load resistance increases, the input ripple to the zener follower:
   (a) decreases;   (b) stays the same;   (c) increases;   (d) none of   ( )
   the foregoing.
3. The pass transistor of Fig. 27-3 has a power dissipation that is closest to:
   (a) 0.25 W;   (b) 0.5 W;   (c) 0.7 W;   (d) 1 W.   ( )
4. The zener diode of Fig. 27-3 has a zener current of approximately:
   (a) 1 mA;   (b) 2.35 mA;   (c) 12.2 mA;   (d) 18.5 mA.
5. In Fig. 27-3, the load current is approximately: ( )
   (a) 1 mA;   (b) 18.5 mA;   (c) 84 mA;   (d) 523 mA.
6. Explain how the regulated power supply of Fig. 27-3 works.

TROUBLESHOOTING

7. Explain why the regulator continues to work even though D_1 is opened.

8. Suppose the circuit of Fig. 27-3 has a collector-emitter short (base, emitter, and collector shorted together). What components are likely to be destroyed?

9. Optional. Instructor's question.

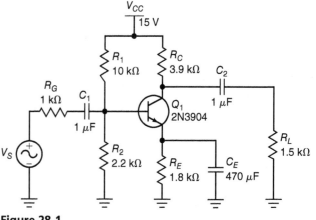

Class-A Amplifiers

In a Class-A amplifier, the transistor operates in the active region throughout the ac cycle. This is equivalent to saying the signal does not drive the transistor into either saturation or cutoff on the ac load line. With a CE amplifier, MPP is the smaller of $2I_{CQ}r_C$ or $2V_{CEQ}$. Some of the important quantities in a Class-A amplifier are the current drain, the maximum transistor power dissipation, the maximum unclipped load power, and the stage efficiency. In this experiment, the voltages, currents, and powers of a Class-A amplifier will be calculated and measured.

GOOD TO KNOW

The Class-A amplifier is often used for audio amplification when a true reproduction of the sound is desired.

Required Reading

Chapter 10 (Secs. 10-1 to 10-3) of *Electronic Principles*, 8th ed.

Equipment

1   signal generator
1   power supply: 15 V
1   transistor: 2N3904
7   ½-W resistors: 220 Ω, 1 kΩ, 1.5 kΩ, 1.8 kΩ, 2.2 kΩ, 3.9 kΩ, 10 kΩ
3   capacitors: two 1 μF, 470 μF (16-V rating or higher)
1   DMM (digital multimeter)
1   oscilloscope

Procedure

CE AMPLIFIER

1. Measure and record the resistor and capacitor values. In Fig. 28-1, calculate the quiescent collector current and the quiescent collector-emitter voltage. Record the answers in Table 28-1.

2. Calculate and record the MPP and the current drain of the stage.
3. Calculate the maximum transistor power dissipation, maximum unclipped load power, dc input power to the stage, and stage efficiency. Record the answers in the theoretical column of Table 28-2.
4. Connect the circuit. Reduce the signal generator to zero. Use the DMM to measure I_{CQ} and V_{CEQ}. Record the data.

Figure 28-1

141

5. Use the oscilloscope to look at the load voltage. Adjust the signal generator until clipping starts on either half-cycle. Notice how the waveform appears squashed on the upper half-cycle and elongated on the lower half-cycle. This nonlinear distortion is being caused by the large changes in r'_e as the collector approaches cutoff and saturation.

6. Reduce the signal generator until the clipping stops. This should be estimated as close as possible because the clipping is soft as the operating point approaches cutoff. Back off enough from the clipping so that the upper peak appears rounded and smooth. Measure and record the peak-to-peak ac voltage. (This measured value is a rough approximation of the MPP value.)

7. Measure and record the total current drain of the stage.

8. Calculate and record the experimental quantities listed in Table 28-2 using the measured data of Table 28-1.

9. Adjust the signal generator to get a peak-to-peak load voltage of 2 V. Notice how much nonlinear distortion there is.

10. Insert a swamping resistor of 220 Ω. Adjust the signal generator to get a load voltage of 2 V_{p-p} and notice how the load signal appears less distorted than before. This improvement in the load signal is explained in the textbook.

TROUBLESHOOTING

11. In Fig. 28-1, assume that R_2 is shorted. Calculate the MPP value and current drain for this trouble. Record in Table 28-3.

12. Repeat Step 11 for each trouble listed in Table 28-3.

13. Build the circuit with each trouble. Measure and record MPP and I_S.

CRITICAL THINKING

14. Select a value of R_E to get maximum MPP value in Fig. 28-1. Record the nearest standard resistance at the top of Table 28-4. Calculate and record the other quantities of Table 28-4.

15. Build the circuit with the design value of R_E. Measure and record MPP and I_S in Table 28-4. Calculate and record the experimental quantities $P_{L(max)}$, P_S, and η using the measured data for MPP and I_S.

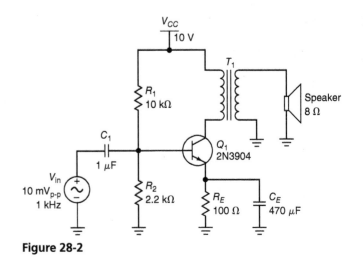

Figure 28-2

APPLICATION (OPTIONAL)

16. Build the circuit shown in Fig. 28-2 using a small 8 Ω speaker connected across the secondary of an audio transformer. The audio transformer should be rated as 600 Ω primary impedance and 8 Ω secondary impedance.

17. Adjust the frequency of the signal generator to 1 kHz. Adjust the output level of the signal generator to produce an ac base voltage of 10 mV_{p-p}. A tone should be heard from the speaker. Vary the frequency and notice that the tone changes.

18. Measure the peak-to-peak collector voltage. It should be from 0.5 to 2 V_{p-p}. Calculate the voltage gain.

19. Measure the peak-to-peak voltage across the speaker. Calculate the speaker power.

ADDITIONAL WORK (OPTIONAL)

20. Have another student insert one of the following troubles into the circuit of Fig. 28-1: open or short any resistor; open any capacitor; open any connecting wire; short the BE, CB, or CE terminals of either transistor. Use only voltage readings of a DMM or an oscilloscope to troubleshoot.

21. Repeat Step 20 several times to build confidence in being able to troubleshoot the circuit for various troubles.

Experiment 28

Lab Partner(s) _____

PARTS USED Nominal Value	Measured Value	PARTS USED Nominal Value	Measured Value	PARTS USED Nominal Value	Measured Value
220 Ω		2.2 kΩ		1 μF	
1 kΩ		3.9 kΩ		1 μF	
1.5 kΩ		10 kΩ		470 μF	
1.8 kΩ					

TABLE 28-1. CE AMPLIFIER

	Calculated	Measured	
		Multisim	Actual
I_{CQ}			
I_{CEQ}			
MPP			
I_S			

TABLE 28-2. POWER AND EFFICIENCY

	Theoretical	Measured	
		Multisim	Actual
$P_{D(max)}$			
$P_{L(max)}$			
P_S			
η			

TABLE 28-3. TROUBLESHOOTING

	Calculated		Measured			
			Multisim		Actual	
Trouble	MPP	I_S	MPP	I_S	MPP	I_S
Shorted R_2						
Open C_E						
Open R_L						
C-E open						

TABLE 28-4. CRITICAL THINKING: R$_E$ = _____

| | Theoretical | Experimental | | | |
		Multisim	% Error	Actual	% Error
MPP					
I_S					
$P_{L(max)}$					
P_S					
η					

Questions for Experiment 28

1. The theoretical MPP value of Fig. 28-1 is approximately: ( )
   (a) 1.1V;   (b) 2.35 V;   (c) 9 V;   (d) 15 V.
2. The total current drain of the amplifier was closest to: ( )
   (a) 1.1 mA;   (b) 2.3 mA;   (c) 4.8 mA;   (d) 6.9 mA.
3. The maximum transistor power dissipation of Fig. 28-1 is approximately: ( )
   (a) 0.46 mW;   (b) 10 mW;   (c) 35.1 mW;   (d) 50 mW.
4. Theoretically, the maximum efficiency of Fig. 28-1 is approximately: ( )
   (a) 0;   (b) 1.3%;   (c) 5%;   (d) 25%.
5. Inserting a swamping resistor in the circuit of Fig. 28-1: ( )
   (a) reduced supply voltage;   (b) increased quiescent collector current;
   (c) decreased nonlinear distortion;   (d) increased ac output compliance.
6. Explain why nonlinear distortion exists in a CE amplifier with a large output signal.

TROUBLESHOOTING

7. What happens to the MPP value when the bypass capacitor CE opens? To the voltage gain?

8. Why did the MPP value increase when R_L was opened?

CRITICAL THINKING

9. Explain how the value of R_E was selected to get maximum MPP.

10. Optional. Instructor's question.

Class-B Push-Pull Amplifiers

In a Class-B push-pull amplifier, each transistor operates in the active region for half of the ac cycle. With a single power supply, the MPP is approximately equal to V_{CC}. Class-B push-pull amplifiers are widely used for the output stage of audio systems because they can deliver more load power than Class-A amplifiers. In fact, the theoretical efficiency of a Class-B push-pull amplifier approaches 78.5 percent, far greater than Class-A.

The main problem with Class-B push-pull amplifiers is setting up a stable Q point near cutoff. The required V_{BE} drop for each transistor is in the vicinity of 0.6 to 0.7 V, with the exact value determined by the quiescent collector current needed to avoid crossover distortion. Since collector current increases by a factor of 10 for each V_{BE} increase of 60 mV, setting up a stable and precise I_{CQ} is difficult. Voltage-divider bias is not practical because the Q point is too sensitive to changes in supply voltage, temperature, and transistor replacement. As discussed in the textbook, there is a real danger of thermal runaway. Diode bias is the usual way to bias a Class-B push-pull amplifier.

In this experiment, 1N4148s (or 1N914s) will be used for the compensating diodes. The complementary transistors used are the 2N3904 (*npn*) and 2N3906 (*pnp*). With the discrete devices of this experiment, the match between the diodes and transistors is only an approximation. For this reason, 10 Ω emitter feedback resistors are used to prevent the excessive collector current that could result from the mismatch between the diodes and the transistors.

GOOD TO KNOW

The pinout for a 2N3906 is the same as the 2N3904.

Emitter Base Collector

Required Reading

Chapter 10 (Secs. 10-4 to 10-7) of *Electronic Principles*, 8th ed.

Equipment

1 signal generator
1 power supply: adjustable from 0 to 15 V
2 diodes: 1N4148 or 1N914
2 transistors: 2N3904, 2N3906
9 ½-W resistors: two 10 Ω, 100 Ω, two 470 Ω, two 680 Ω, two 4.7 kΩ

3 capacitors: two 1 μF, 100-μF (16-V rating or higher)
2 DMMs (digital multimeters)
1 oscilloscope

Procedure

CROSSOVER DISTORTION

1. Measure and record the resistor and capacitor values. Adjust the power supply to 5 V and then build the circuit of Fig. 29-1*a* with this supply voltage.

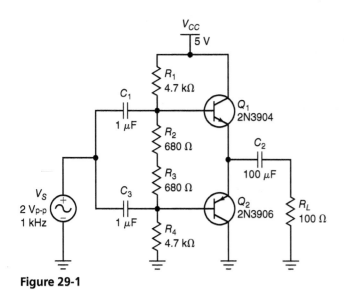

Figure 29-1

2. Set the input frequency to 1 kHz and the signal level across the signal generator at 2 V_{p-p}.

3. Observe the output signal across the load resistor (100 Ω). Crossover distortion should be visible on the oscilloscope.

SINGLE-SUPPLY CIRCUIT

4. Build the circuit of Fig. 29-2. Reduce the output of the signal generator to 0 V. Measure the current through the diodes. Record this measurement in Table 29-1.

5. Measure and record the collector current in the upper 2N3904. The diode current and transistor currents of

Table 29-1 may differ substantially. The main reason is the mismatch in characteristics between discrete diodes and transistors. With integrated circuits, these two currents would be equal to a close approximation.

6. Adjust the output of the signal generator to 2 V_{p-p}. Record this value in Table 29-1.

7. Measure and record the peak-to-peak output voltage across the 100 Ω load resistor.

8. Increase the output level of the signal generator until clipping just starts on the final output signal. This is the approximate value of the MPP. Record the MPP in Table 29-1.

9. Repeat Steps 4 through 8 for two more pairs of transistors. If Multisim is being used, change the transistors' beta values to 150 and 200, respectively.

DUAL-SUPPLY CIRCUIT

10. Build the circuit of Fig. 29-3. Reduce the output of the signal generator to 0 V. Measure the current through the diodes. Record this measurement in Table 29-2.

11. Measure and record the collector current in the upper 2N3904.

12. Adjust the output of the signal generator to 2 V_{p-p}. Record this value in Table 29-2.

13. Measure and record the peak-to-peak output voltage across the 100 Ω load resistor.

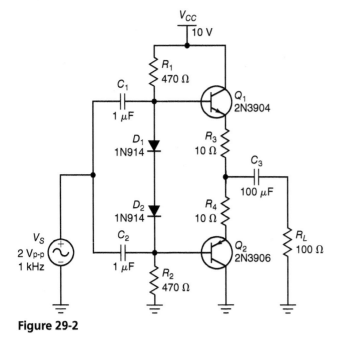

Figure 29-2

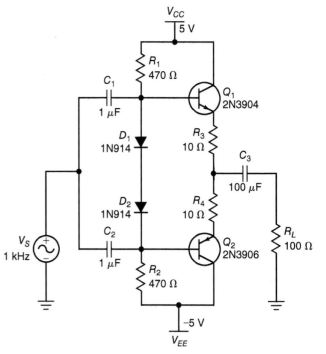

Figure 29-3

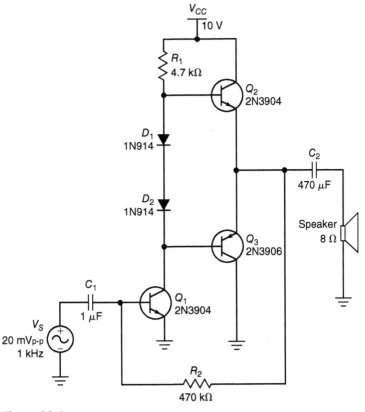

Figure 29-4

14. Increase the output level of the signal generator until clipping just starts on the final output signal. This is the approximate value of the MPP. Record the MPP in Table 29-2.

15. Repeat Steps 10 through 14 for two more pairs of transistors. If Multisim is being used, change the transistors' beta values to 150 and 200, respectively.

APPLICATION (OPTIONAL)

16. Connect the circuit shown in Fig. 29-4 using a small 8 Ω speaker. Adjust the frequency of the signal generator to 2 kHz. Adjust the output level of the signal generator to produce an ac input voltage of 100 mV$_{\text{p-p}}$. A tone should be heard from the speaker. Vary the frequency and notice that the tone changes.

Experiment 29

Lab Partner(s) _____

PARTS USED Nominal Value	Measured Value	PARTS USED Nominal Value	Measured Value	PARTS USED Nominal Value	Measured Value
10 Ω		470 Ω		4.7 kΩ	
10 Ω		680 Ω		1 μF	
100 Ω		680 Ω		1 μF	
470 Ω		4.7 kΩ		100 μF	

TABLE 29-1. SINGLE-SUPPLY CIRCUIT

Transistors	Multisim					Actual				
	I_{diode}	I_C	v_{in}	v_{out}	MPP	I_{diode}	I_C	v_{in}	v_{out}	MPP
1										
2										
3										

TABLE 29-2. DUAL-SUPPLY CIRCUIT

Transistors	Multisim					Actual				
	I_{diode}	I_C	v_{in}	v_{out}	MPP	I_{diode}	I_C	v_{in}	v_{out}	MPP
1										
2										
3										

Questions for Experiment 29

1. In Fig. 29-1, crossover distortion will occur when the supply voltage is: ( )
   (a) negative; (b) too low; (c) too high; (d) unstable.
2. Crossover distortion can be reduced by having:
   (a) low supply voltage; (b) two *npn* transistors; (c) a small
   quiescent I_C; (d) no load resistor.
3. In Fig. 29-2, the 10 Ω resistors produce: ( )
   (a) negative feedback; (b) positive feedback; (c) crossover distortion;
   (d) more MPP.
4. In Fig. 29-2, the 1N4148 (or 1N914) diodes are: ( )
   (a) germanium; (b) zeners; (c) forward-biased; (d) good matches
   for the transistors.
5. In Fig. 29-3, the dc voltage between the 10-Ω resistors ideally is: ( )
   (a) 0; (b) 5 V; (c) −5 V; (d) 10 V.
6. In Fig. 29-4, the maximum speaker power is measured in: ( )
   (a) nanowatts; (b) microwatts; (c) milliwatts; (d) watts.

7. Explain why the diode and transistor currents are different in Table 29-1.

CRITICAL THINKING

8. Explain how the diodes in Fig. 29-2 can help prevent thermal runaway by the transistors.

9. Optional. Instructor's question.

An Audio Amplifier

In this experiment, a discrete audio amplifier with a Class-A input stage and a Class-B push-pull emitter follower, as shown in Fig. 30-1, will be constructed and analyzed. Adjustment R_7 is included to allow the Q_3 emitter voltage to be set at $+5$ V, half the supply voltage. This lets the output signal swing equally in both directions to get maximum MPP value. Some of the resistance values have not been optimized, allowing for improvements to the design (optional). The circuit is fairly complicated, so don't rush when wiring it. Double-check all connections before applying power.

GOOD TO KNOW

Some signal generators have difficulty producing a small amplitude signal. A voltage divider can be used to obtain very small amplitude signals.

Required Reading

Chapters 9 and 10 of *Electronic Principles,* 8th ed.

Equipment

1 signal generator
1 power supply: 10 V
2 diodes: 1N4148 or 1N914
4 transistors: three 2N3904, 2N3906
9 ½-W resistors: two 100 Ω, two 1 kΩ, 2.2 kΩ, 3.9 kΩ, 4.7 kΩ, two 10 kΩ
1 potentiometer: 5 kΩ
4 capacitors: two 1 μF, 100 μF, 470 μF (16-V rating or higher)
1 DMM (digital multimeter)
1 oscilloscope

Procedure

AUDIO AMPLIFIER

1. Measure and record the resistor and capacitor values. Assume that the base resistance of Q_2 is adjusted to produce a quiescent voltage of $+5$ V at the emitter of Q_3. Calculate and record all dc voltages listed in Table 30-1.

2. Assume an ac load voltage of 4 $V_{p\text{-}p}$. Also assume all h_{fe} values are 100. Calculate and record all ac voltages listed in Table 30-2.

3. Build the audio amplifier of Fig. 30-1. Adjust the base resistor of Q_2 to produce a dc voltage $+5$ V at the emitter of Q_3.

4. Set the input frequency to 1 kHz and adjust the amplitude of the signal generator to produce an ac load voltage of 4 $V_{p\text{-}p}$.

5. Use the DMM to measure all dc voltages listed in Table 30-1. Use the oscilloscope to measure all ac voltages in Table 30-2. Record the data.

TROUBLESHOOTING

6. Ask another student to insert the following trouble into the circuit: open any component or connecting wire.

7. Locate and repair the trouble.

8. Repeat Steps 6 and 7 two or more times as desired.

CRITICAL THINKING

9. Measure the MPP value. Record it as the initial value in Table 30-3.

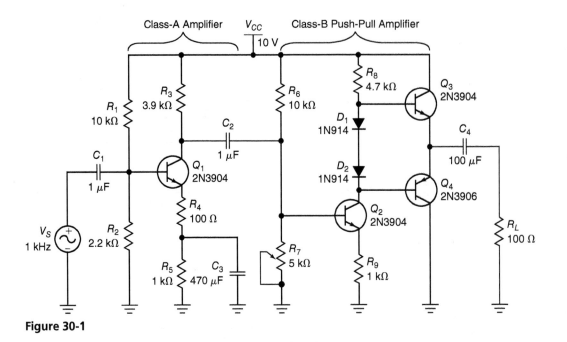

Figure 30-1

10. Try to figure out how to increase the MPP value. For instance, changing certain resistances will increase the MPP value.
11. Insert the changes and measure the MPP value.
12. When the MPP value has been increased as much as possible, record the final MPP in Table 30-3. Also record the changes that were inserted.

APPLICATION (OPTIONAL)

13. Replace the 100 Ω load with an 8 Ω speaker. Adjust the output level of the signal generator to produce an audible tone. Vary the frequency and notice that the tone changes.

ADDITIONAL WORK (OPTIONAL)

14. Write an essay on how to troubleshoot a circuit like Fig. 30-1. Discuss at least three different troubles and the methods that can be used to isolate the trouble.

Experiment 30

Lab Partner(s) _____

PARTS USED Nominal Value	Measured Value	PARTS USED Nominal Value	Measured Value	PARTS USED Nominal Value	Measured Value	PARTS USED Nominal Value	Measured Value
100 Ω		2.2 kΩ		10 kΩ		1 μF	
100 Ω		3.9 kΩ				1 μF	
1 kΩ		4.7 kΩ				100 μF	
1 kΩ		10 kΩ				470 μF	

TABLE 30-1. DC VOLTAGES

	Calculated			Measured					
				Multisim			Actual		
	B	E	C	B	E	C	B	E	C
Q_1									
Q_2									
Q_3									
Q_4									

TABLE 30-2. AC VOLTAGES

	Calculated			Measured					
				Multisim			Actual		
	B	E	C	B	E	C	B	E	C
Q_1									
Q_2									
Q_3									
Q_4									

TABLE 30-3. CRITICAL THINKING: INITIAL MPP = _____

	Final MPP	
	Multisim	Actual
Final MPP		
Changes were as follows:		

Questions for Experiment 30

1. Resistor R_7 is adjusted to get approximately: (   )
   (a) 10 mA through R_8;   (b) +10 V at the collector of Q_3;   (c) +5 V at the
   emitter of Q_3;   (d) 0 V at the collector of Q_4.
2. The capacitive reactance of 100 μF at 1 kHz is approximately: (   )
   (a) 1.59 Ω;   (b) 6.28 Ω;   (c) 100 Ω;   (d) 1 kΩ.
3. The amplifier of Fig. 30-1 had a voltage gain closest to: (   )
   (a) 1;   (b) 25;   (c) 100;   (d) 200.
4. The measured ac voltage at the base of Q_3 was slightly higher than the ac output   (   )
   voltage because:
   (a) the output capacitor was too small;   (b) of the drop across r'_e;
   (c) R_7 was adjustable;   (d) Q_2 was an *npn* transistor.
5. One reason the MPP value of the output stage is less than V_{CC} is because of: (   )
   (a) voltage drop across X_C;   (b) offset voltage of the 1N914s;   (c) V_{BE}
   drops of the output transistors;   (d) power dissipation by the load resistor.
6. Explain how the audio amplifier of Fig. 30-1 works.

TROUBLESHOOTING

7. Suppose a 1N4148 (or 1N914) shorts in Fig. 30-1. What kind of symptoms will this produce?

8. Resistor R_4 of Fig. 30-1 is shorted by a solder bridge. Describe some of the symptoms that result.

CRITICAL THINKING

9. Why is the MPP value less than V_{CC} in Fig. 30-1?

10. Why did the changes made to Fig. 30-1 increase the MPP?

Class-C Amplifiers

In a Class-C amplifier, the transistor operates in the active region for less than 180° of the ac cycle. Typically, the conduction angle is much smaller than 180° and the collector current is a train of narrow pulses. This highly nonsinusoidal current contains a fundamental frequency plus harmonics. A tuned Class-C amplifier has a resonant tank circuit that is tuned to the fundamental frequency. This produces a sinusoidal output voltage of frequency f_r. In a frequency multiplier, the resonant tank circuit is tuned to the nth harmonic so that the sinusoidal output has a frequency of nf_r.

In this experiment, a tuned Class-C amplifier and a frequency multiplier will be built.

GOOD TO KNOW

When dc coupled, the oscilloscope will display the ac waveform along with any dc level the ac waveform is riding on. When ac coupled, only the ac waveform is displayed, with the dc level removed.

Required Reading

Chapter 10 (Sec. 10-8) of *Electronic Principles,* 8th ed.

Equipment

1   RF generator
1   power supply: 10 V
1   transistor: 2N3904
3   ½-W resistors: 220 Ω, two 100 kΩ
1   inductor: 33 mH (or smaller value as close as possible)
3   capacitors: two 1 μF, 470 pF
1   DMM (digital multimeter)
1   oscilloscope
1   frequency counter (optional)

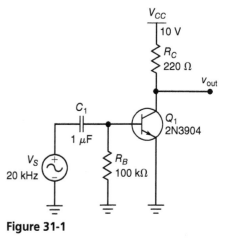

Figure 31-1

Procedure

1. Measure and record the inductor, resistor, and capacitor values. Build the circuit shown in Fig. 31-1.
2. Set the input frequency of 20 kHz. Adjust the signal level to get narrow output pulses with a peak-to-peak value of 6 V.
3. Measure the pulse width W and period T. Record these values in Table 31-1. Calculate and record the duty cycle.
4. Look at the signal on the base. The waveform should be negatively clamped. With the oscilloscope on dc coupled, measure and record the positive and negative peak voltages of this clamped signal.

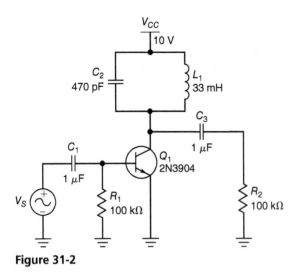

Figure 31-2

5. In Fig. 31-2, calculate the resonant frequency of the tuned amplifier. Record the answer in Table 31-2.

6. Assume that the Q_L of the coil is 15. Calculate and record the other quantities listed in Table 31-2.

7. Build the circuit. Use the oscilloscope to look at the base signal. Set the RF generator to produce a 2-V_{p-p}, 40 kHz sine wave.

8. Use the oscilloscope to look at the collector signal. Vary the input frequency until the output signal reaches a maximum value (resonance).

9. Adjust the signal level as needed to get 15 V_{p-p} at the collector.

10. Repeat Steps 8 and 9 until the circuit is resonant with an output of 15 V_{p-p}. (This repetitive adjustment of frequency and signal level is called "rocking in.")

RESONANT FREQUENCY, BANDWIDTH, AND CIRCUIT Q

11. Use the frequency counter to measure the resonant frequency. If a frequency counter is not available, use the oscilloscope or generator indication. Record in Table 31-3.

12. Measure f_1 and f_2. Calculate and record the bandwidth in Table 31-3.

13. Readjust the input frequency to get resonance. Calculate the circuit Q using the f_r and B of Table 31-3. Record the circuit Q.

MPP, CURRENT DRAIN, AND DC CLAMPING

14. Increase the input signal level until the output signal just starts clipping. Back off slightly from this level until the signal stops clipping. Record the MPP in Table 31-3.

15. Use the DMM as an ammeter to measure the current drain of the circuit. Record I_S.

FREQUENCY MULTIPLIER

16. Reduce the input frequency and notice that the output signal decreases (off resonance). Continue decreasing the input frequency until the output signal again reaches a maximum value (resonant to a harmonic). Use the frequency counter or oscilloscope to measure the input and output frequencies. Record in Table 31-4. Divide f_{out} by f_{in}, round the answer off to the nearest integer, and notice that it equals 2. The circuit is now operating as a X2 frequency multiplier.

17. Again decrease the input frequency until you find another resonance. Measure and record the input and output frequencies. This time, the f_{out}/f_{in} ratio should be approximately 3. The circuit is now acting like a X3 frequency multiplier.

TROUBLESHOOTING (OPTIONAL)

18. In Fig. 31-2, assume that R_1 is open. Estimate the ac load voltage for this trouble. Record in Table 31-5.

19. Repeat Step 18 for each trouble listed in Table 31-5.

20. Build the circuit with each trouble. Measure and record the ac load voltage.

CRITICAL THINKING

21. Select a value of C_2 (nearest standard value) to get a resonant frequency of approximately 250 kHz.

22. Build the circuit with the design value of C_2. Tune to the fundamental frequency. Record the capacitance and frequency here:

$C_2 =$ _____ .

$f_r =$ _____ .

ADDITIONAL WORK (OPTIONAL)

23. When testing RF circuits with an oscilloscope, one must be constantly aware of the loading effect of the probes. An inexpensive oscilloscope may add as much as 200 pF to a circuit on the X1 probe position. If you use the X10 probe position, the input capacitance decreases to 20 pF. Therefore, use the X10 position whenever possible.

24. Probe capacitances will depend on the quality of the oscilloscope being used. Find out what the probe capacitances are for the X1 and X10 positions of the oscilloscope being used. Record the values:

$C_{(X1)} =$ _____ .

$C_{(X10)} =$ _____ .

25. Replace the 33 mH inductor with a 330 μH inductor (or an inductor near this value). Since the resonant frequency is inversely proportional to the square root of inductance, the new resonant frequency will be approximately 400 kHz.

26. Look at the output with the X1 position of the probe. Measure the resonant frequency and record the value:

$f_r =$ _____ (X1)

27. Switch the probe to the X10 position. Increase the sensitivity by a factor of 10 to compensate for the probe attenuation of X10.

28. Repeat Step 26 and record the resonant frequency:

$f_r =$ _____ (X10)

29. The frequency in Step 28 should be slightly higher than the frequency in Step 26 because the probe will be adding less capacitance to the circuit during the measurement of resonant frequency.

30. When no probe is connected, is the resonant frequency higher, lower, or the same as the frequency in Step 28?

Answer = _____.

31. What was learned about the effect that an oscilloscope probe has on a circuit under test conditions?

32. If all the wires were shortened as much as possible in the circuit, what effect would this have on the resonant frequency? Explain the reason for the answer given.

Experiment 31

Lab Partner(s) _____

PARTS USED Nominal Value	Measured Value	PARTS USED Nominal Value	Measured Value
220 Ω		1 μF	
100 kΩ		1 μF	
100 kΩ		470 pF	
33 mH			

TABLE 31-1. WAVEFORMS

	W	T	D	+peak	−peak
Multisim					
Actual					

TABLE 31-2. CALCULATIONS FOR TUNED AMPLIFIER

f_r	X_L	R_S	R_P	r_c	Q	B	MPP	$P_{L(max)}$

TABLE 31-3. MEASUREMENTS FOR TUNED AMPLIFIER

	f_r	B	Q	MPP	I_S
Multisim					
Actual					

TABLE 31-4. FREQUENCY MULTIPLIER

	f_{in}	f_{out}	n
Multisim			2
			3
Actual			2
			3

TABLE 31-5. TROUBLESHOOTING

Trouble	Estimated v_{out}	Measured v_{out}	
		Multisim	Actual
R_1 open			
Q_1 C-E open			
C_2 shorted			
C_3 open			

Questions for Experiment 31

1. The duty cycle of Table 31-1 is closest to: ( )
   (a) 1%;   (b) 10%;   (c) 31.6%;   (d) 75%.
2. The MPP of Table 31-2 is approximately: ( )
   (a) 0.7 V;   (b) 1.4 V;   (c) 10 V;   (d) 20 V.
3. To calculate the total dc input power to the tuned amplifier, the   ( )
   measured current drain of Table 31-3 can be multiplied by the:
   (a) MPP;   (b) circuit Q;   (c) supply voltage;   (d) bandwidth.
4. Assume the current drain equals 0.25 mA in Fig. 31-2. If the ac output compliance   ( )
   is 19.6 V, then the stage efficiency is approximately:
   (a) 5%;   (b) 19%;   (c) 47%;   (d) 73%.
5. When f_{in} = 4.75 kHz and f_{out} = 19 kHz, a frequency multiplier is tuned to which   ( )
   harmonic of the fundamental frequency?
   (a) first;   (b) second;   (c) third;   (d) fourth.
6. Briefly explain how a tuned Class-C amplifier works.

7. Explain how a frequency multiplier works.

TROUBLESHOOTING

8. The input voltage driving the tuned Class-C amplifier of Fig. 31-2 is 1 V_{p-p}. The output voltage is zero. Name the most likely trouble.

CRITICAL THINKING

9. In this experiment, the tuned Class-C amplifier has a stage efficiency of only 20 percent, more or less, depending on the components used. What was the cause of this poor efficiency? Explain the answer.

10. Optional. Instructor's question.

Multistage Transistor Application

Dc wall adapter transformers are commonly used to power electronic circuits. In this experiment, a multistage audio amplifier with diode input protection will be constructed from discrete components and powered by a 12- to 15-volt DC wall adapter transformer. A variable power supply can be substituted if needed. The multistage transistor circuit uses a diode protection circuit (D_1 and D_2) to prevent the amplifier from being damaged from too large of an input signal. The first stage (Q_1) is composed of a Class-A amplifier. The Class-A amplifier provides distortion-free amplification. The second stage (Q_2) is composed of a common emitter (CE) driver. The CE driver behaves as a swamped voltage amplifier. The final stage (Q_3 and Q_4) is a Class-B push-pull amplifier. The compensating diodes (D_3 and D_3) prevent thermal runaway while providing the biasing voltage for the emitter diodes in Q_3 and Q_4.

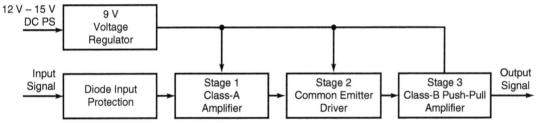

Figure SA3-1

In this experiment, the use of an oscilloscope to trace a signal through multiple stages of the circuit will be introduced. This experiment will also provide an opportunity to troubleshoot a multistage circuit.

GOOD TO KNOW

When dc coupled, the oscilloscope will display the ac waveform and any dc level the waveform is riding on. When ac coupled, only the ac waveform is displayed, with any dc level removed.

Required Reading

Chapters 6 to 10 of *Electronic Principles,* 8th ed.

Equipment

1   variable dc power supply ($+12$ V to $+15$ V)
1   voltage regulator: LM7809 (or equivalent)
4   transistors: three 2N3604, one 2N3606

4   diodes: 1N914 or 1N4148
10 ½-W resistors: two 100 Ω, 220 Ω two 1 kΩ, 2.2 kΩ, 4.7 kΩ, 5.1 kΩ, two 10 kΩ
7   capacitors: two 0.1 μF, 0.33 μF, two 1 μF, 100 μF, 470 μF (25-V rating or higher)
1   5 kΩ potentiometer
1   DMM (digital multimeter)
1   oscilloscope

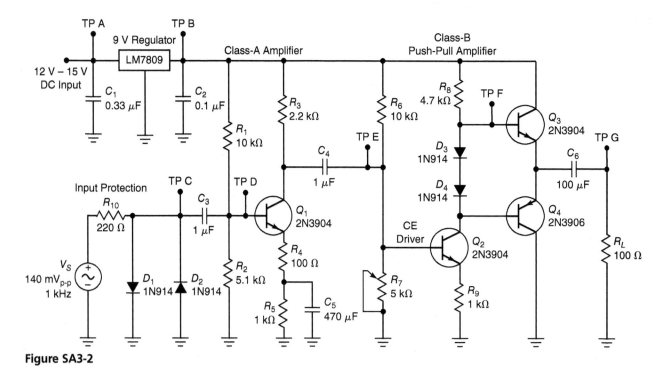

Figure SA3-2

Procedure

MULTISTAGE CIRCUIT ANALYSIS

1. Sketch the pinout of the LM7809 9V voltage regulator. Using a DMM, measure and record the value of the resistors and capacitors. Change the DMM mode to Diode Test, and measure and record the voltage across both the silicon diodes when forward and reverse biased. Assume that the base resistance of Q_2 is adjusted to produce a quiescent voltage of +5.5 V at the emitter of Q_3 (TP F). Calculate and record all dc voltages listed in Table SA3-1.

2. Assume an ac load voltage of 4 V_{p-p}. Also, assume all h_{fe} values are 100. Calculate and record all ac voltages listed in Table SA3-2.

3. Build the multistage circuit, as shown in Fig. SA3-2.

4. Use a wall adapter transformer with a dc output between 12 and 15 V or a variable dc power supply set to 13 V.

5. Using a DMM, adjust R_7 for a +5.2 V dc voltage at test point F. This value will change slightly when adjustments are made in the next step. (*Note:* All voltage measurements are with respect to ground.)

6. While observing the signal at test point G, set the input frequency to 1 kHz and adjust the amplitude of

the signal generator to produce an ac load voltage of 4 V_{p-p} (TP G). While observing the waveform at test point G, adjust R_7 until the clipping is removed.

7. Based on the material in the textbook and the data sheet for the LM7809, predict the voltages that would be present at test points A and B. Record the predictions in Table SA3-3.

8. Using the oscilloscope, observe and measure the input dc voltage signal at test point A with channel 1; observe and measure the dc voltage and ac peak-to-peak ripple voltage at test point B with channel 2. Record the measured values in Table SA3-3. (*Note:* Use dc coupling for the dc voltage measurement and ac coupling for the ac ripple voltage measurement.)

9. Based on the material in the textbook, predict the dc and ac voltages that would be present at test points C, E, and G. Record the predictions in Table SA3-3.

10. Using the oscilloscope, observe and measure the dc voltage and ac peak-to-peak signal voltage at test point C with channel 1. Using the oscilloscope, observe and measure the dc voltage and ac peak-to-peak signal voltage present at test point E with channel 2. Record the measured values in Table SA3-3. (*Note:* Use dc coupling for the dc voltage measurement and ac coupling for the ac ripple voltage measurement.)

11. Using the oscilloscope, observe and measure the dc voltage and ac peak-to-peak signal voltage at test point C with channel 1. Using the oscilloscope, observe and measure the dc voltage and ac peak-to-peak signal voltage present at test point G with channel 2. Record the measured values in Table SA3-3. (*Note:* Use dc coupling for the dc voltage measurement and ac coupling for the ac ripple voltage measurement.)

TROUBLESHOOTING

12. As electrolytic capacitors age, their actual capacitance value tends to decrease. Simulate a failing capacitor in Fig. SA3-2 by replacing C_3 with a 0.1 μF capacitor. Predict the dc and ac voltages that would be present at test points C and D. Record the predictions in Table SA3-4. Using the oscilloscope, while observing the input signal at test point C with channel 1, measure the ac peak-to-peak signal voltage at test point D with channel 2. Record the measured value in Table SA3-4 and explain why the signal was smaller at test point D than at test point C.

13. Further simulate a failing capacitor in Fig. SA3-2 by removing C_3. Predict the dc and ac voltages that would be present at test points C and D. Record the predictions in Table SA3-4. Using the oscilloscope, while observing the input signal at test point C with channel 1, measure the ac peak-to-peak signal voltage at test point D with channel 2. Record the measured value in Table SA3-4 and explain what happened to the signal.

14. Q_1 has a partially bypassed emitter resistor. How would the gain of the Class-A amplifier be affected if C_5 failed as an open?

CRITICAL THINKING

15. The diode input protection circuit prohibits large input signals from entering the Class-A amplifier. Diodes D_1 and D_2 will appear as an open to the input signal below the diodes' forward-biased voltage. Once the input signal exceeds this voltage, the diodes conduct, limiting the input signal to their forward-biased voltage. To view this process, the gain of the amplifier must be reduced. Remove C_5 from the circuit. Calculate the new voltage gain for the Class-A amplifier with C_5 removed. Record the calculated value in Table SA4-5.

16. Using the oscilloscope, while observing the input signal at test point D with channel 1, measure the ac peak-to-peak signal voltage at test point E with channel 2. Calculate the measured voltage gain. Record the calculated value in Table SA4-5.

17. While observing the input signal at test point C with channel 1 and the output signal at test point G with channel 2, increase the amplitude of the signal generator until the diodes conduct. The diodes conducting will cause a square wave to be present at test point C. Continue to increase the amplitude of the input signal to 4 V_{p-p} while observing the output signal at test point G. Does the amplitude of the output signal change? Should the amplitude of the output signal increase? Support this conclusion with observed circuit behavior.

System Application 3

Lab Partner(s) _____

PARTS USED Nominal Value	Measured Value		PARTS USED Nominal Value	Measured Value
100 Ω			10 kΩ	
100 Ω			10 kΩ	
220 Ω			0.1 μF	
1 kΩ			0.1 μF	
1 kΩ			0.33 μF	
2.2 kΩ			1 μF	
4.7 kΩ			1 μF	
5.1 kΩ			100 μF	
	V_F	V_R	470 μF	
1N914			LM7809 PINOUT	
1N914				
1N914				
1N914				

TABLE SA3-1. DC VOLTAGE MEASUREMENTS

	Calculated			Measured					
				Multisim			Actual		
	B	E	C	B	E	C	B	E	C
Q_1									
Q_2									
Q_3									
Q_4									

TABLE SA3-2. AC VOLTAGE MEASUREMENTS

	Calculated			Measured					
				Multisim			Actual		
	B	E	C	B	E	C	B	E	C
Q_1									
Q_2									
Q_3									
Q_4									

TABLE SA3-3. VOLTAGE MEASUREMENTS

Test Point	Predicted Value		Multisim		Actual	
	DC	AC	DC	AC	DC	AC
A						
B						
C						
E						
G						

TABLE SA3-4. VOLTAGE MEASUREMENTS

Test Point $C_3 = 0.1 \mu F$	Predicted Value		Multisim		Actual	
	DC	AC	DC	AC	DC	AC
C						
D						
C_3 Rmvd	DC	AC	DC	AC	DC	AC
C						
D						

Explain the reason for the smaller signal at test point D as compared to test point C when C_3 was 0.1 μF:

Explain what happened to the signal at test point D when C_3 was removed:

TABLE SA3-5. VOLTAGE GAIN MEASUREMENTS

	Calculated	Multisim	Actual
	A_V	A_V	A_V
C_5 Removed			

Questions for System Application 3

1. R_7 is adjusted to get _____ at test point F.
   (a) +9 V;   (b) +3.2 V;   (c) +5.2 V;   (d) +7.8 V.
2. According to the data sheet for the LM7809, a minimum of _____ is required to obtain +9 V on the output.
   (a) +15 V;   (b) +11.5 V;   (c) +10 V;   (d) +13.8 V.
3. Diodes D_1 and D_2 limit the input signal to approximately _____.
   (a) +0.7 V;   (b) +1.4 V;   (c) +100 mV;   (d) +11.5 V.
4. The r'$_e$ value for Q_1 is _____.
   (a) 17 Ω;   (b) 31 Ω;   (c) 5 Ω;   (d) 11.7 Ω.

5. The r'_e value for Q_2 is _____.
   (a) 17 Ω;   (b) 31 Ω;   (c) 5 Ω;   (d) 11.7 Ω.
6. The Class-A amplifier has a voltage gain (A_V) of _____ when C_5 is present.
   (a) 7;   (b) 1;   (c) 22;   (d) 18.
7. The Class-A amplifier has a voltage gain (A_V) of _____ when C_5 is removed.
   (a) 7;   (b) 1;   (c) 22;   (d) 18.
8. The value of X_{C1} increased to _____ when C_1 was 0.1 μF.
   (a) 1951 Ω;   (b) 159 Ω;   (c) 1591 Ω;   (d) 591 Ω.
9. A Class-A amplifier has _____ distortion.
   (a) low;   (b) high;   (c) intermittent;   (d) no.
10. Diodes D_3 and D_3 prevent _____.
   (a) thermal runaway;   (b) voltage spikes;   (c) oscillations;   (d) current fluctuation.

32

JFET Bias

Gate bias is the simplest but worst way to bias a JFET for linear operation because the drain current depends on the exact value of V_{GS}. Since V_{GS} has a large variation, the drain current has a large variation. Self-bias offers some improvement because the source resistor produces local feedback, which reduces the effect of V_{GS}. When large supply voltages are available, voltage-divider bias results in a relatively stable Q point. Finally, current-source bias can produce the most stable Q point because a bipolar current source sets up the drain current through the JFET.

In this experiment, the JFET will be biased in the different methods just described. This will illustrate the stability of each type of bias.

GOOD TO KNOW

Be careful when connecting the JFET to the circuit. The gate is not the center pin, but the end pin, as seen in the image to the right.

Drain Source Gate

Required Reading

Chapter 11 (Secs. 11-1 to 11-5) of *Electronic Principles,* 8th ed.

Equipment

2   power supplies: adjustable from 0 to ±15 V
1   transistor: 2N3904
3   JFETs: MPF102 (or equivalent)
7   ½-W resistors: 470 Ω, 680 Ω, 1 kΩ, 2.2 kΩ, 6.8 kΩ, 33 kΩ, 100 kΩ
1   DMM (digital multimeter)

Procedure

MEASURING I_{DSS}

1. Measure and record the resistor values. Locate a data sheet for an MPF102 on the Internet. Notice that $V_{GS(off)}$ has a maximum of -8 V; the minimum value

is not specified. Also notice that I_{DSS} has a minimum value of 2 mA and a maximum value of 20 mA.

2. Build the circuit of Fig. 32-1*a*. Measure the drain current. Record this value of I_{DSS} in Table 32-1. (*Note:* Because of heating effects, the drain current may decrease slowly. Take the reading as soon after power-up as possible.)

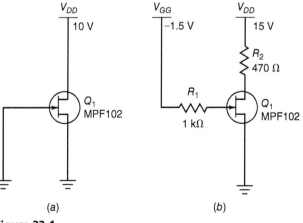

Figure 32-1

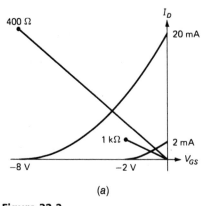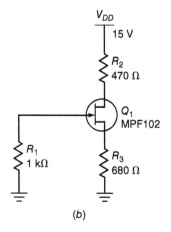

(a)   (b)

Figure 32-2

3. Repeat Step 2 for the other JFETs. If Multisim is being used, replace the MPF102 JFET with 2N5485 and 2N5486 JFETs for this step.

GATE BIAS

4. With gate bias, apply a fixed gate voltage that reverse-biases the gate of the JFET. This produces a drain current that is less than I_{DSS}. The problem is that the drain current cannot be accurately predicted for mass production because of the variation in the required V_{GS}. The following steps will illustrate this point.
5. Build the circuit of Fig. 32-1b. Measure V_{GS}, I_D, and V_D. Record the data in Table 32-2.
6. Repeat Step 5 for the other JFETs. In a random set of three JFETs, the drain current usually will show a significant variation from one JFET to another.

MEASURING $V_{GS(OFF)}$

7. Here is an approximate way to measure $V_{GS(off)}$. Insert the first JFET into the gate-biased circuit. Increase the negative gate supply voltage of Fig. 32-1b until the drain current is approximately 1 μA. (If the DMM cannot measure down to 1 μA, then use 10 μA or 100 μA, or whatever low value the instructor indicates.) Record the approximate $V_{GS(off)}$ in Table 32-1.
8. Repeat Step 7 for the other two JFETs.

SELF-BIAS

9. The data sheet of a JFET lists a maximum I_{DSS} of 20 mA and a maximum $V_{GS(off)}$ of −8 V. If a JFET has these values, its transconductance curve appears, as shown in Fig. 32-2a. Notice that the approximate source resistance for self-bias is

$$R_S = \frac{8\text{ V}}{20\text{ mA}} = 400\ \Omega$$

The minimum I_{DSS} is 2 mA. For this experiment, assume that the minimum $V_{GS(off)}$ is −2 V. A JFET with these values has the lower transconductance curve

of Fig. 32-2a, and required R_S is 1 kΩ. An average source resistance is around 700 Ω, so a 680 Ω resistor will be used in the test circuit.

10. Assume a V_{GS} of −2 V in Fig. 32-2b. Calculate I_D, V_D, and V_S. Record the results in Table 32-3.
11. Build the self-bias circuit of Fig. 32-2b. Measure I_D, V_D, and V_S. Record the data in Table 32-3.
12. Repeat Step 11 for the other JFETs.
13. Notice that the drain current of the self-biased circuit (Table 32-3) has less variation than the drain current of the gate-biased circuit (Table 32-2).

VOLTAGE-DIVIDER BIAS

14. Assume that V_{GS} is −2 V in Fig. 32-3. Calculate I_D, V_D, and V_S. Record the results in Table 32-4.
15. Build the circuit. Measure and record the quantities of Table 32-4.
16. Repeat Step 15 for the other JFETs. Notice that the drain current of Table 33-4 has less variation than the drain currents of Tables 32-2 and 32-3.

CURRENT-SOURCE BIAS

17. Assume V_{GS} is −2 V in Fig. 32-4. Calculate and record the quantities of Table 32-5.
18. Build the circuit. Measure I_D, V_D, and V_S. Record the results.

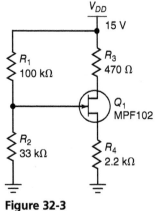

Figure 32-3

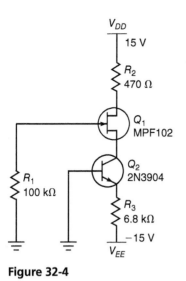

Figure 32-4

TROUBLESHOOTING

19. Assume R_2 is shorted in Fig. 32-4. Calculate and record V_D in Table 32-6.

20. Insert the trouble into the circuit. Measure and record V_D.

21. Repeat Steps 19 and 20 for the other troubles listed in Table 32-6.

CRITICAL THINKING

22. Use Eq. (11-1) from the textbook and the data of Table 32-1 to calculate the source resistance of a self-biased circuit. Average the three source resistances. Record the answer at the top of Table 32-7.

23. Build the circuit of Fig. 32-2*b* with the design value of R_S. Measure I_D, V_D, and V_S. Record the data.

24. Repeat Step 23 for the other JFETs.

ADDITIONAL WORK (OPTIONAL)

25. Measure V_{GS}, V_{DS}, and I_D and then graph the drain curves.

26. In Fig. 32-1*a*, measure and record I_D for different values of V_{DS}.

27. Remove the short on the gate. Apply a V_{GS} of -1 V and repeat Step 26.

28. Apply several more negative values of V_{GS} and repeat Step 26 until there is enough data for a graph like Fig. 11-5 in the textbook.

29. Graph I_D versus V_{DS} for different values of V_{GS}.

Experiment 32

Lab Partner(s) _____

PARTS USED Nominal Value	Measured Value	PARTS USED Nominal Value	Measured Value
470 Ω		6.8 kΩ	
680 Ω		33 kΩ	
1 kΩ		100 kΩ	
2.2 kΩ			

TABLE 32-1. JFET DATA

	Multisim		Actual	
JFET	I_{DSS}	$V_{GS(off)}$	I_{DSS}	$V_{GS(off)}$
1				
2				
3				

TABLE 32-2. GATE BIAS: $V_{GG} = -1.5$ V

	Multisim			Actual		
JFET	V_{GS}	I_D	V_D	V_{GS}	I_D	V_D
1						
2						
3						

TABLE 32-3. SELF-BIAS

	Calculated			Measured					
				Multisim			Actual		
JFET	I_D	V_D	V_S	I_D	V_D	V_S	I_D	V_D	V_S
1									
2									
3									

TABLE 32-4. VOLTAGE-DIVIDER BIAS

	Calculated			Measured					
				Multisim			Actual		
JFET	I_D	V_D	V_S	I_D	V_D	V_S	I_D	V_D	V_S
1									
2									
3									

TABLE 32-5. CURRENT-SOURCE BIAS

| JFET | Calculated | | | Measured | | | | | |
| | | | | Multisim | | | Actual | | |
	I_D	V_D	V_S	I_D	V_D	V_S	I_D	V_D	V_S
1									
2									
3									

TABLE 32-6. TROUBLESHOOTING

| Trouble | Estimated | | Measured | | | |
| | | | Multisim | | Actual | |
	V_D	V_C	V_D	V_C	V_D	V_C
R_2 shorted						
Q_2 C-E shorted						
R_3 open						

TABLE 32-7. DESIGN: R_S = _____

| JFET | Multisim | | | | Actual | | | |
	I_S	V_D	V_S	% Error	I_S	V_D	V_S	% Error
1								
2								
3								

Questions for Experiment 32

1. This experiment proved that the circuit with the least variation in drain current was:　(　)
   (a) gate bias;　　(b) self-bias;　　(c) voltage-feedback;　　(d) current-source bias.
2. With gate bias, the drain current has which of the following:　　　　　　(　)
   (a) almost constant value;　　(b) constant drain voltage;　　(c) values greater than I_{DSS};　　(d) large variations.
3. Self-bias is better than:　　　　　　　　　　　　　　　　　　　(　)
   (a) gate bias;　　(b) voltage-divider bias;　　(c) current-source bias;
   (d) emitter bias.
4. The voltage across the source resistor of Fig. 32-3 equals the gate voltage plus the　(　)
   magnitude of:
   (a) V_{GS};　　(b) V_D;　　(c) V_S;　　(d) V_{DD}.
5. The voltage-divider bias of Fig. 32-3 would be more stable if the designer:　　(　)
   (a) decreased the supply voltage;　　(b) increased the supply voltage;
   (c) decreased R_2;　　(d) increased R_1.
6. Briefly discuss the variations in drain current for the four types of JFET bias.

TROUBLESHOOTING

7. Suppose the voltage across the source resistor of Fig. 32-3 is zero. Name three possible troubles.

8. The drain voltage of Fig. 32-2 is 15 V. Name three possible troubles.

CRITICAL THINKING

9. What value of R_S was used in the design? How was this value determined?

10. Optional. Instructor's question.

33

JFET Amplifiers

Because the transconductance curve of a JFET is parabolic, large-signal operation of a CS amplifier produces square-law distortion. This is why a CS amplifier is usually operated small-signal. JFET amplifiers cannot compete with bipolar amplifiers when it comes to voltage gain. Because g_m is relatively low, the typical CS amplifier has a relatively low voltage gain.

The CD amplifier, better known as the source follower, is analogous to the emitter follower. The voltage gain approaches unity and the input impedance approaches infinity, limited only by the external biasing resistors connected to the gate. The source follower is a popular circuit often found near the front end of measuring instruments.

In this experiment, a CS amplifier and a source follower will be built and analyzed. The experimental data will support the topics covered in the textbook.

GOOD TO KNOW

Be careful when connecting the JFET to the circuit. The gate is not the center pin but the end pin, as seen in the image to the right.

Drain Source Gate

Required Reading

Chapter 11 (Secs. 11-6 and 11-7) of *Electronic Principles,* 8th ed.

Equipment

1   signal generator
1   power supply: 15 V
3   JFETs: MPF102 (or equivalent)
4   ½-W resistors: 1 kΩ, two 2.2 kΩ, 220 kΩ
3   capacitors: two 1 μF, 100 μF (16-V rating or higher)
1   potentiometer: 5 kΩ
1   oscilloscope
1   DMM (digital multimeter)

Procedure

CS AMPLIFIER

1. Assume that the JFET of Fig. 33-1 has a typical g_m of 2000 μS. Calculate the unloaded voltage gain, output voltage, and output impedance. Record the results in Table 33-1.

2. Measure and record the resistor and capacitor values. Build the circuit with R_L equal to infinity (no load resistor).

3. Adjust the signal generator to 1 kHz. Set the signal level to 0.1 V_{p-p} across the input.

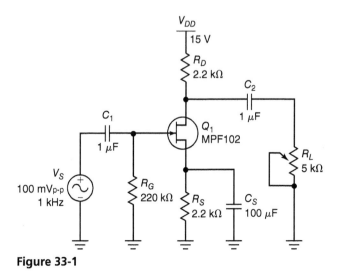

Figure 33-1

4. Observe the output signal with an oscilloscope. It should be an amplified sine wave. Measure and record the peak-to-peak output voltage. Then calculate the voltage gain. Record the answer as the measured A_V in Table 33-1.

5. Connect the 5 kΩ potentiometer as a variable load resistance. Adjust this load resistance until the output voltage is half of the unloaded output voltage.

6. Disconnect the 5 kΩ potentiometer and measure its resistance. Record this as r_{out} in Table 33-1. (*Note:* The Thevenin or output impedance was just experimentally determined by the matched-load method.)

7. Repeat Steps 1 through 6 for the other JFETs. If Multisim is being used, replace the MPF102 JFET with 2N5485 and 2N5486 JFETs for this step.

SOURCE FOLLOWER

8. Assume a typical g_m of 2000 μS in Fig. 33-2. Calculate the unloaded voltage gain, output voltage, and output impedance. Record the results in Table 33-2.

9. Build the circuit with R_L equal to infinity. Adjust the frequency to 1 kHz and the signal level to 1 $V_{p\text{-}p}$ across the input.

10. Measure and record the output voltage. Calculate the voltage gain and record as the measured A_V in Table 33-2.

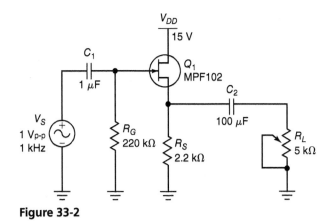

Figure 33-2

11. Measure and record the output impedance by the matched-load method used earlier.

12. Repeat Steps 8 through 11 for the other JFETs.

TROUBLESHOOTING

13. Table 33-3 lists dc and ac symptoms for Fig. 33-2. Try to figure out what trouble would produce these symptoms. When the suspected trouble is determined, insert the trouble into the circuit. Then measure the dc and ac voltages to verify that the suspected trouble is the actual trouble that is causing the symptoms. Record the trouble in Table 33-3.

14. Repeat Step 13 for the other symptoms listed in Table 33-3.

CRITICAL THINKING

15. Redesign the source follower of Fig. 33-2 so that it uses voltage-divider bias. Assume V_{GS} is −2 V and select R_1 and R_2 to produce a V_S of +7.5 V. Record the resistance values at the top of Table 33-4.

16. Build the redesigned source follower. Measure the dc voltage at the source. Record V_S in Table 33-4. Also, set the ac input voltage to 1 $V_{p\text{-}p}$. Measure the ac output voltage. Calculate and record the unloaded voltage gain.

17. Repeat Step 16 for each JFET.

ADDITIONAL WORK (OPTIONAL)

18. The effects different values of V_{GS} have on the voltage gain and output impedance will be explored in this section.

19. Change the source resistor in Fig. 33-1 from 2.2 kΩ to 470 Ω. Measure the voltage gain and output impedance of the circuit.

20. Change the source resistor in Fig. 33-2 from 2.2 kΩ to 470 Ω. Measure the voltage gain and output impedance of the circuit.

21. What assumptions can be made from the results of Steps 19 and 20?

Experiment 33

Lab Partner(s) _____

PARTS USED Nominal Value	Measured Value	PARTS USED Nominal Value	Measured Value
1 kΩ		1 μF	
2.2 kΩ		1 μF	
2.2 kΩ		100 μF	
220 kΩ			

TABLE 33-1. CS AMPLIFIER

	Calculated			Measured					
				Multisim			Actual		
JFET	v_{out}	A_V	r_{out}	v_{out}	A_V	r_{out}	v_{out}	A_V	r_{out}
1									
2									
3									

TABLE 33-2. SOURCE FOLLOWER

	Calculated			Measured					
				Multisim			Actual		
JFET	v_{out}	A_V	r_{out}	v_{out}	A_V	r_{out}	v_{out}	A_V	r_{out}
1									
2									
3									

TABLE 33-3. TROUBLESHOOTING

DC Symptoms			AC Symptoms				
V_G	V_D	V_S	v_g	v_d	v_s	v_{out}	Trouble
0	15 V	3.7 V	1 V	0	0	0	
0	15 V	3.7 V	1 V	0	0.82 V	0	
0	15 V	3.7 V	0	0	0	0	
0	0	0	1 V	0	0	0	

TABLE 33-4. CRITICAL THINKING: R_1 = _____; R_2 = _____

	Multisim			Actual		
JFET	V_S	A_V	% Error	V_S	A_V	% Error
1						
2						
3						

Questions for Experiment 33

1. The calculated voltage gain of Table 33-1 is approximately: ( )
   (a) 0.44; (b) 1; (c) 4.4; (d) 9.4.

2. The output impedance of Fig. 33-1 is closest to: ( )
   (a) 407 Ω; (b) 2.2 kΩ; (c) 5 kΩ; (d) 220 kΩ.

3. The voltage gain of the source follower was closest to: ( )
   (a) 0.5; (b) 0.8; (c) 1; (d) 4.4.

4. The source follower had an output impedance closest to: ( )
   (a) 0; (b) 100 Ω; (c) 200 Ω; (d) 400 Ω.

5. The main advantage of a JFET amplifier is its: ( )
   (a) high voltage gain; (b) low drain current; (c) high input impedance;
   (d) high transconductance.

6. Compare the voltage gain of a CS amplifier like Fig. 33-1 to a bipolar junction CE amplifier.

TROUBLESHOOTING

7. In Fig. 33-2, all dc voltages are normal. The ac gate voltage and source voltage are normal. There is no output voltage. What is the most likely trouble?

8. All dc voltages are normal in Fig. 33-2. All ac voltages are zero. What is the most likely trouble?

CRITICAL THINKING

9. How were the values of R_1 and R_2 determined?

10. Optional. Instructor's question.

JFET Applications

One of the main applications of JFETs is the analog switch. In this application, a JFET acts either like an open switch or like a closed switch. This allows the building of circuits that either transmit an ac signal or block it from the output terminals.

In the ohmic region, a JFET acts like a voltage-variable resistance instead of a current source. The value of $r_{ds(on)}$ can be changed by changing V_{GS}. When a JFET is used as a voltage-variable resistance, the ac signal should be small, typically less than 100 mV.

Another application of the JFET is with automatic gain control (AGC). Because the g_m of a JFET varies with the Q point, amplifiers can be built where voltage gain is controlled by an AGC voltage.

In this experiment, various JFET circuits will be built and examined for a better understanding as to how the JFET can be used as a switch, voltage-variable resistance, and AGC device.

GOOD TO KNOW

The data sheets for all the semiconductor devices used in this lab manual can be found on the Internet by searching for their respective part numbers.

Required Reading

Chapter 11 (Secs. 11-8 and 11-9) of *Electronic Principles*, 8th ed.

Equipment

1   signal generator
2   power supplies: adjustable to ± 15 V
1   diode: 1N4148 or 1N914
3   JFETs: MPF102 (or equivalent)
4   ½-W resistors: 2.2 kΩ, 10 kΩ, two 100 kΩ
2   capacitors: 1 μF
1   switch: SPST
1   DMM (digital multimeter)
1   oscilloscope

Procedure

ANALOG SWITCH

1. Measure and record the resistor and capacitor values. Locate a data sheet for an MPF102 on the Internet. Measure the approximate R_{DS} of each JFET as follows. Short the gate and source together. Connect the positive lead of an ohmmeter to the drain and the negative lead to the source. Record the values of R_{DS} in Table 34-1. Throughout this experiment, R_{DS} may be used as an approximation for r_{ds}.
2. In Fig. 34-1, calculate v_{out} for each JFET when V_{GS} is zero. Record the results in Table 34-1.
3. Build the circuit with the ac signal source at 100 mV$_{p-p}$ and 1 kHz.

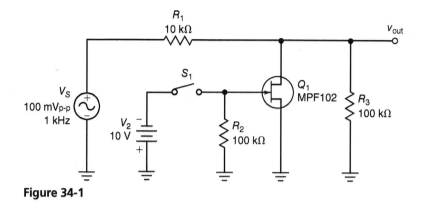

Figure 34-1

4. Measure and record the ac output voltage with S_1 open and S_1 closed.

5. Repeat Step 4 for the other two JFETs. If Multisim is being used, replace the MPF102 JFET with 2N5485 and 2N5486 JFETs for this step.

JFET CHOPPER

6. Predict what the output voltage of Fig. 34-2 should look like. Sketch the expected waveform in Table 34-2.

7. Build the circuit with the specified ac input voltages and frequencies.

8. View the output voltage on an oscilloscope. Set the trigger select of the oscilloscope to the 1 kHz signal. Slowly vary the 1 kHz frequency until a steady chopped waveform is displayed on the oscilloscope. Sketch the waveform in Table 34-2.

VOLTAGE-VARIABLE RESISTANCE

9. As V_{GG} is varied from zero to a value more negative than $V_{GS(off)}$, the peak-to-peak output voltage of Fig. 34-3 will change. Should the peak-to-peak voltage increase or decrease?

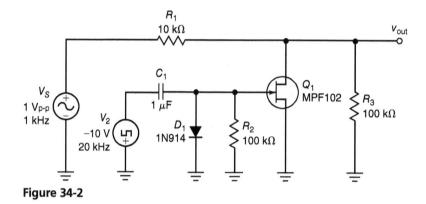

Figure 34-2

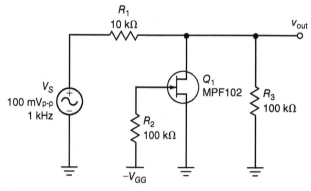

Figure 34-3

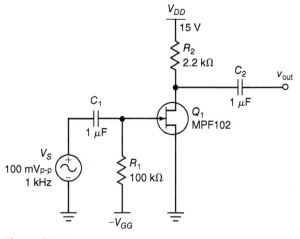

Figure 34-4

10. Build the circuit.
11. Set up each value of V_{GG} shown in Table 34-3. Measure and record the ac output voltage. Calculate and record the value of r_{ds}.

AGC CIRCUIT

12. Build the circuit of Fig. 34-4. Adjust V_{GG} to get the maximum output signal. Measure and record v_{out} and V_{GG} in Table 34-4.
13. Adjust V_{GG} until v_{out} drops in half. Measure and record v_{out} and V_{GG}.
14. Repeat Step 13 two times.

TROUBLESHOOTING

15. Assume that V_{GG} is set to produce the maximum output in Fig. 34-4. For each set of dc and ac symptoms listed in Table 34-5, figure out what the corresponding trouble may be. Insert the suspected trouble, then check the dc and ac voltages. When each trouble is confirmed, record it in Table 34-5.

CRITICAL THINKING

16. Select a value of R_1 in Fig. 34-1 that increases the attenuation by a factor of 10 when S_1 is open. Connect and test the circuit with the design value.

Experiment 34

Lab Partner(s) _____

PARTS USED Nominal Value	Measured Value	PARTS USED Nominal Value	Measured Value
2.2 kΩ		100 kΩ	
10 kΩ		1 μF	
100 kΩ		1 μF	

TABLE 34-1. JFET ANALOG SWITCH

			Measured v_{out}			
			Multisim		Actual	
JFET	Measured R_{DS}	Calculated v_{out}	S_1 open	S_1 closed	S_1 open	S_1 closed
1						
2						
3						

TABLE 34-2. JFET CHOPPER

Expected Waveform (sketch below)	Experimental Waveform

TABLE 34-3. VOLTAGE-VARIABLE RESISTANCE DATA

	Multisim		Actual	
V_{GG}	v_{out}	r_{ds}	v_{out}	r_{ds}
0				
−0.5 V				
−1 V				
−1.5 V				
−2 V				
−2.5 V				
−3 V				
−3.5 V				
−4 V				
−4.5 V				
−5 V				

TABLE 34-4. AGC CIRCUIT

Condition	Multisim		Actual	
	v_{out}	V_{GG}	v_{out}	V_{GG}
Maximum output				
Max/2				
Max/4				
Max/8				

TABLE 34-5. TROUBLESHOOTING

DC Symptoms		AC Symptoms		Trouble
V_G	V_D	v_g	v_d	
OK	OK	0	0	
OK	0	OK	0	
0	0	0	0	

Questions for Experiment 34

1. The JFET analog switch of Fig. 34-1 attenuates the signal when: ( )
   (a) it is on the negative half-cycles; (b) S_1 is open; (c) S_1 is closed; (d) R_1 is shorted.

2. The gate circuit of Fig. 34-2 contains a: ( )
   (a) 1 kHz frequency; (b) positive dc clamper; (c) negative dc clamper; (d) forward bias.

3. With the voltage-variable resistance of Fig. 34-3, the output signal increases when V_{GS}: ( )
   (a) becomes more negative; (b) stays the same; (c) becomes more positive; (d) is zero.

4. In the AGC circuit of Fig. 34-4, the output voltage decreases when: ( )
   (a) the ac input signal increases; (b) V_{GG} goes to zero; (c) V_{GG} becomes more negative; (d) none of the foregoing.

5. Explain how the JFET analog switch of Fig. 34-1 works.

6. Explain how the voltage-variable resistance circuit of Fig. 34-3 works.

TROUBLESHOOTING

7. The dc drain voltage is zero in Fig. 34-4. Name three possible troubles.

8. The output voltage of Fig. 34-1 is zero with S_1 open or closed. Name three possible troubles.

CRITICAL THINKING

9. Which is better for a JFET analog switch: a low or high $r_{ds(on)}$ when $V_{GS} = 0$? Explain the reasoning behind the answer given.

10. Optional. Instructor's question.

Power FETs

The MOSFET has an insulated gate that results in extremely high input impedance. The depletion-type MOSFET, also called a normally on MOSFET, can operate in either the depletion or enhancement mode. The enhancement-type MOSFET, also called a normally off MOSFET, can operate only in the enhancement mode. Power FETs are enhancement-type MOSFETs that are useful in applications requiring high load power, including audio amplifiers, RF amplfiers, and interfacing. In this experiment, circuits containing power FETS will be built and analyzed.

GOOD TO KNOW

The matched-load method is often used to determine the output impedance of an amplifier: Measure the no-load output voltage. Add a potentiometer as the load, and vary the load resistance until the load voltage is half of the no-load voltage. The load resistance is equal to the Thevenin or output impedance.

Required Reading

Chapter 12 (Secs. 12-4 to 12-9) of *Electronic Principles,* 8th ed.

Equipment

1   signal generator
1   power supply: 15 V
1   diode: L53RD or similar red LED
1   power FET: IRF510 or VN10KM
5   ½-W resistors: 330 Ω, 560 Ω, 1 kΩ, two 22 kΩ
2   capacitors: 0.1 μF, 100 μF (16-V rating or higher)
1   switch: SPST
2   potentiometers: 1 kΩ, 5 kΩ (or similar low-resistance pots)
1   DMM (digital multimeter)
1   oscilloscope
Graph paper

Procedure

THRESHOLD VOLTAGE

1. Measure and record the resistor and capacitor values. Build the circuit shown in Fig. 35-1a. (See Fig. 35-1b for connections.)
2. Most data sheets define threshold voltage as the gate voltage that produces a drain current of 10 μA. Adjust the coarse and fine controls to get a drain current of 10 μA.
3. Record the threshold voltage in Table 35-1.

TRANSCONDUCTANCE CURVE

4. Set the fine control to maximum. Adjust the coarse control to produce a drain current slightly more than 10 μA. Adjust the fine control to get each drain current listed in Table 35-2. Record each gate voltage.
5. Graph the data, I_D versus V_{GS}.

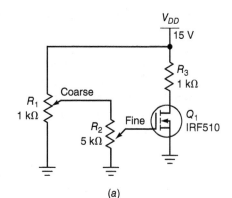

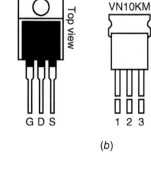

(a)

(b)

Figure 35-1

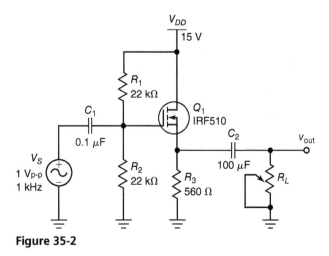

Figure 35-2

VOLTAGE-DIVIDER BIAS

6. Assume a V_{GS} of +2 V in Fig. 35-2. Calculate I_D, V_G, and V_S. Record the results in Table 35-3.
7. Build the circuit.
8. Measure and record I_D, V_G, and V_S.

SOURCE FOLLOWER

9. Calculate the g_m of the transistor in Fig. 35-2. (Use the graph from Step 5 for this.) Calculate the voltage gain and output impedance for a load resistance of infinity (R_L open). Record the results in Table 35-4.
10. Build the circuit with an infinite R_L. Measure and record the voltage gain for an input of 1 V_{p-p}. Also measure and record the MPP value.
11. Measure and record the output impedance using the matched-load method (described in Experiment 33).

DRIVING AN LED

12. Assume the R_{DS} of the power FET is 5 Ω. Calculate the I_D and V_D in Fig. 35-3 with S_1 closed. Record the results in Table 35-5.
13. Build the circuit. Open and close S_1. The LED should light when S_1 is closed and go out when S_1 is open.
14. Close S_1. Measure I_D and V_D.
15. Record the results in Table 35-5.

TROUBLESHOOTING

16. Table 35-6 lists some symptoms for the circuit of Fig. 35-3. Try to figure out the corresponding trouble. Insert the suspected trouble and verify the symptoms. Record each trouble.

CRITICAL THINKING

17. Select a value of current-limiting resistance in Fig. 35-3 that produces an LED current of approximately 20 mA.
18. Connect the circuit with the design value. Measure the LED current. Record the data here:

    $R = $ _____

    $I = $ _____

APPLICATION (OPTIONAL)—POWER FET-CONTROLLED MOTOR

19. Build the circuit of Fig. 35-4 using a dc motor such as the MCM 28-12810 (used earlier in Experiment 7). Do not apply power at this time.
20. Firmly attach a piece of electrical tape to the motor shaft, leaving a ½-inch flap. The movement of the tape will indicate the direction of rotation at slower speeds. Now, apply the +12 V of supply voltage. If the motor is running, vary the potentiometer until it stops.

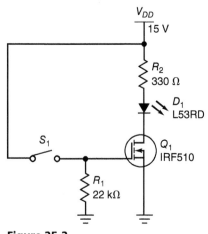

Figure 35-3

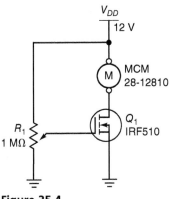

Figure 35-4

21. Measure V_{GS} while varying the 1-MΩ potentiometer. Record the voltages at which the motor turns on and it turns off.
22. Slowly vary the potentiometer until the motor turns slowly enough to identify the direction of rotation.
23. Reverse the leads on the motor. Again, slowly vary the potentiometer until the motor turns slowly enough to identify the direction of rotation.
24. What was learned in Steps 21 through 23?

MOISTURE-SENSING CIRCUIT

25. Build the circuit of Fig. 35-5.

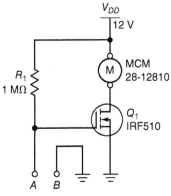

Figure 35-5

26. The motor will run as soon as power is applied to the circuit.
27. Short points A and B. The motor will stop. Open the points, and the motor will run again.
28. If a small container of water is available, dip points A and B into the water. The motor will stop. Remove the points from the water, and the motor will run again. What was learned in Steps 26 through 28?

SOFT TURN-ON CIRCUIT

29. Build the circuit of Fig. 35-6. The motor should not be running when the switch is open.
30. Calculate and record the charging time constant when the switch is closed:

_____

31. Connect a voltmeter across the capacitor. Then, close the switch and watch the voltmeter as it measures the increasing voltage across the capacitor. At some point the motor will turn on.
32. Remove the power from the circuit. Replace the 10 μF with 100 μF. Then, repeat Steps 30 and 31.
33. What was learned in Steps 30 through 32?

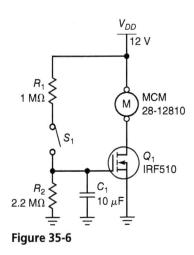

Figure 35-6

Experiment 35

Lab Partner(s) _____

PARTS USED Nominal Value	Measured Value	PARTS USED Nominal Value	Measured Value
330 Ω		22 kΩ	
560 Ω		0.1 μF	
1 kΩ		100 μF	
22 kΩ			

TABLE 35-1. THRESHOLD VOLTAGE

	Multisim	Actual
$V_{GS(\text{th})}$		

TABLE 35-2. TRANSCONDUCTANCE CURVE

I_D	V_{GS} Multisim	V_{GS} Actual
10 μA		
0.5 mA		
1 mA		
2 mA		
3 mA		
4 mA		
5 mA		
6 mA		
7 mA		
8 mA		
9 mA		
10 mA		

TABLE 35-3. VOLTAGE-DIVIDER BIAS

Calculated			Measured					
			Multisim			Actual		
I_D	V_G	V_S	I_D	V_G	V_S	I_D	V_G	V_S

TABLE 35-4. SOURCE FOLLOWER

Calculated			Measured					
			Multisim			Actual		
g_m	A_V	r_{out}	A_V	MPP	r_{out}	A_V	MPP	r_{out}

TABLE 35-5. DRIVING A LOAD

Calculated		Measured			
		Multisim		Actual	
I_D	V_D	I_D	V_D	I_D	V_D

TABLE 35-6. TROUBLESHOOTING

S_1	V_D	LED	Trouble
Closed	0 V	Out	
Closed	$\approx +14$ V	Out	
Open	0 V	On	

Questions for Experiment 35

1. The threshold voltage of the power FET was the gate voltage that produced a drain current of: ( )
   (a) 10 μA; (b) 100 μA; (c) 1 mA; (d) 10 mA.
2. When the drain current is approximately 10 mA, the transconductance of the power FET used in this experiment was closest to: ( )
   (a) 100 μS; (b) 1000 μS; (c) 2500 μS; (d) 30,000 μS.
3. The calculated drain current in Table 35-3 is approximately: ( )
   (a) 1.1 mA; (b) 5.26 mA; (c) 9.82 mA; (d) 12.5 mA.
4. The voltage gain in Table 35-4 is: ( )
   (a) less than 0.5; (b) slightly less than unity; (c) greater than unity; (d) around 10.
5. The measured drain voltage of Table 35-5 was: ( )
   (a) large; (b) less than 1 V; (c) equal to supply voltage; (d) unstable.
6. Describe what the power FET does in Fig. 35-3.

TROUBLESHOOTING

7. What is the last trouble recorded in Table 35-6? Why does it produce the given symptoms?

8. Suppose the dc voltage across the 560 Ω resistor of Fig. 35-2 is zero. Name three possible troubles.

CRITICAL THINKING

9. What value of resistance was recorded in Step 18? How was this resistance determined?

36

The Silicon Controlled Rectifier

The silicon controlled rectifier (SCR) acts like a normally off switch. To turn it on, a trigger voltage and current are applied to the gate. Once on, the SCR acts like a closed switch. The trigger can be removed and the SCR will continue to conduct. The only way to stop the SCR from conducting is to reduce the current flowing through the SCR to below the holding current rating of the SCR.

GOOD TO KNOW

Due to the low trigger voltage and current requirements of an SCR, they are often used as an interface between a digital circuit and an analog circuit.

Required Reading

Chapter 13 (Secs. 13-1 to 13-4) of *Electronic Principles*, 8th ed.

Equipment

1 power supply: adjustable from approximately 0 to 15 V with current limiting
1 red LED: L53RD or equivalent
2 transistors: 2N3904, 2N3906
1 SCR: 2N6505 or equivalent
6 ½-W resistors: two 330 Ω, 470 Ω, two 1 kΩ, 10 kΩ
1 potentiometer: 1 kΩ
1 DMM (digital multimeter)
1 oscilloscope

Procedure

TRANSISTOR LATCH

1. Measure and record the resistor and capacitor values. The transistor latch of Fig. 36-1 simulates an SCR. Assume that the LED of Fig. 36-1 is off. For a V_{CC} of +15 V, calculate the voltage between point A and ground. Record in Table 36-1. Also calculate and record the LED current.

2. Assume the switch of Fig. 36-1 is momentarily closed, then opened. Calculate and record the voltage at point A to ground for a V_{CC} of +15 V. Also calculate and record the LED current.

3. Build the circuit with the switch open and a V_{CC} of +15 V.

4. The LED should be out. If not, reduce the supply voltage to zero, then back to +15 V.

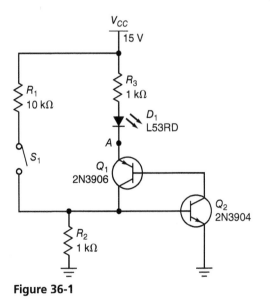

Figure 36-1

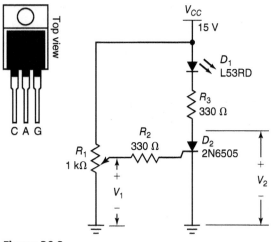

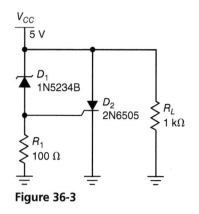

Figure 36-3

Figure 36-2

5. With the LED off, measure and record the voltage at point A and the LED current.

6. Close the switch. The LED should come on.

7. Open the switch. The LED should stay on.

8. With the LED on, measure and record the voltage at point A. Measure and record the LED current. (When the circuit is opened to insert the ammeter, the LED will go out. Close the switch to turn on the LED and measure the current.)

9. With the LED on, reduce the supply voltage until the LED goes off. Then increase the supply voltage and notice that the LED remains off.

10. Close the switch, then open it. The LED should be on.

SCR CIRCUIT

11. In Fig. 36-2, assume that V_{CC} is $+15$ V and that the LED is off. Calculate and record V_2 in Table 36-2. Also calculate and record the LED current.

12. Assume the LED is on. Calculate and record V_2 and I_{LED}.

13. Assume a gate trigger current of 7 mA and a gate trigger voltage of 0.75 V. If the LED is off, what is the value of V_1 needed to turn on the LED? Record the results in Table 36-2.

14. Assume a holding current of 6 mA. Calculate and record the value of V_{CC} where SCR turns off.

15. Build the circuit with R_1 reduced to zero.

16. Adjust V_{CC} to $+15$ V. The LED should now be out. (If the LED is on, reduce V_{CC} to zero, then increase it back to $+15$ V. The LED should now be out.) Measure and record V_2 and I_{LED}.

17. Slowly increase V_1 until the LED just comes on. Measure and record V_2 and I_{LED}. Also measure and record V_1.

18. Reduce V_1 to zero. Slowly decrease V_{CC} until the LED just goes out. Measure and record V_{CC}.

APPLICATION (OPTIONAL)

19. Build the crowbar circuit of Fig. 36-3.

20. Connect a dc voltmeter across the 1 kΩ. It should read approximately 5 V.

21. Slowly increase the supply voltage while watching the voltmeter reading. The crowbar should activate somewhere above 6.2 V. When it does, the load voltage will drop to a low value.

22. Reduce the supply voltage to zero. Then, repeat Step 21. What was learned about a crowbar?

23. What was learned in Steps 21 and 22?

24. Build the motor-control circuit of Fig. 36-4 using a small dc motor such as the MCM 28-12810 (9 to 18 V). Do not apply power at this time.

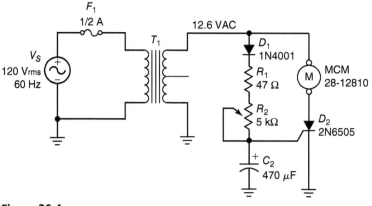

Figure 36-4

194

25. Firmly attach a piece of electrical tape to the motor shaft, leaving a ½-inch flap.

26. Now, apply power and vary the 5 kΩ resistance in both directions over its entire range. The motor speed should change from maximum to minimum, and vice versa. (*Note:* If the motor cannot be turned off, add 4.7 kΩ in series with the variable resistance.)

27. Vary the resistance slowly and notice that the speed of the motor can be controlled from very slow to full speed.

28. Reverse the motor leads and repeat Steps 26 and 27.

29. What was learned in Steps 26 through 28?

30. In Fig. 36-4, replace the dc motor with a 1 kΩ resistor. Connect an oscilloscope to measure the voltage across the anode-to-cathode of the SCR.

31. Vary the resistance of the 5-kΩ potentiometer while observing the SCR's changing firing angle. Adjust the potentiometer until the conduction angle is approximately 135°.

Experiment 36

Lab Partner(s) _____

PARTS USED Nominal Value	Measured Value	PARTS USED Nominal Value	Measured Value	PARTS USED Nominal Value	Measured Value
330 Ω		470 Ω		1 kΩ	
330 Ω		1 kΩ		10 kΩ	

TABLE 36-1. TRANSISTOR LATCH

	Calculated		Measured			
			Multisim		Actual	
	V_A	I_{LED}	V_A	I_{LED}	V_A	I_{LED}
LED						
Off						
On						

TABLE 36-2. SCR CIRCUIT

	Calculated		Measured			
			Multisim		Actual	
	V_2	I_{LED}	V_2	I_{LED}	V_2	I_{LED}
LED off						
LED on						
Triggering I_{GT}						
Holding I_H						

Questions for Experiment 36

1. When the LED of Fig. 36-1 is on, the voltage from point A to ground is closest to: ( )
   (a) 0; (b) 3 V; (c) 10 V; (d) V_{CC}.

2. After the LED of Fig. 36-1 comes on, the current through it is approximately: ( )
   (a) 0; (b) 5 mA; (c) 9 mA; (d) 13 mA.

3. When the switch of Fig. 36-1 is closed, the voltage across the lower 1 kΩ is: ( )
   (a) 0; (b) 0.7 V; (c) 1.5 V; (d) 15 V.

4. When the switch of Fig. 36-1 is closed, the current through the 2N3904 is closest to: ( )
   (a) 0; (b) 5 mA; (c) 8 mA; (d) 14 mA.

5. After the LED of Fig. 36-2 comes on, the only way to make it go off is to: ( )
   (a) reduce V_1 to zero; (b) increase V_1 to 15 V; (c) reduce V_{CC} toward zero; (d) increase V_{CC} to 15 V.

6. The calculated V_1 of Table 36-2 needed for triggering is closest to: ( )
   (a) 1 V; (b) 1.9 V; (c) 3.06 V; (d) 4.78 V.

7. Assume an LED voltage of 2 V in Fig. 36-2. The current through the SCR when it   ( )
   is conducting is closest to:
   (a) 0;   (b) 20 mA;   (c) 40 mA;   (d) 75 mA.
8. When V_{CC} is 15 V and the LED is off in Fig. 36-2, V_2 is equal to: ( )
   (a) 1 V;   (b) 1.9 V;   (c) 3.06 V;   (d) 15 V.
9. What is meant by the term "latch" as it applies to the SCR?

10. After an SCR is turned on and anode current is flowing, how is it turned off?

11. Optional. Instructor's question.

Frequency Effects

An ac amplifier is normally operated in some middle range of frequencies referred to as midband, where its voltage gain is approximately constant. Below the midband the response is down 3 dB at the lower cutoff frequency. Above the midband the response is down 3 dB at the upper cutoff frequency. The -3 dB point occurs at 0.707 of $v_{Midband}$. The cutoff frequency is sometimes called the corner frequency.

Decibels are useful for measuring and specifying the frequency response of amplifiers. When using Bode plots, the y-axis represents the voltage gain in decibels and the x-axis represents the frequency range. The x-axis is in logarithmic scale, allowing the frequency response to be viewed over multiple decades. Most spreadsheet programs have the ability to have one or both axes in logarithmic scale.

Risetime is a useful way to specify the response of an amplifier. As discussed in the textbook, the risetime of an amplifier can be calculated and the result used to calculate its upper cutoff frequency by using Eq. (14-29) in the textbook.

GOOD TO KNOW

Voltage gain is often expressed in decibels. The formula to convert A_V to $A_{V(dB)}$ is:
$$A_{V(dB)} = 20 \log (A_V).$$

Required Reading

Chapter 14 of *Electronic Principles*, 8th ed.

Equipment

1 function generator
1 power supply: adjustable to 10 V
1 DMM (digital multimeter)
1 transistor: 2N3904
5 ½-W resistors: 1 kΩ, 2.2 kΩ, 3.9 kΩ, 8.2 kΩ, 10 kΩ
6 capacitors: 220 pF, 470 pF, 0.1 μF, 0.47 μF, 10 μF, 470 μF
1 oscilloscope

Procedure

1. Measure and record the resistor and capacitor values. Answer the following questions for capacitors C_1 to C_4 in Fig. 37-1.

Which are the capacitors that influence the lower cutoff frequency?

Answer = _____.

Which is the capacitor that affects the upper cutoff frequency?

Answer = _____.

2. Build the circuit shown in Fig. 37-1 using the following capacitances: $C_1 = 0.1$ μF, $C_2 = 10$ μF, $C_3 = 470$ μF, and $C_4 = 470$ pF. Capacitance C_4 is included to show the effects of stray wiring and other capacitances across the load.

3. Set the frequency of the input sine wave to 10 kHz. Use channel 1 of the oscilloscope to measure peak-to-peak voltage across the function generator (adjust to 20 mV$_{p-p}$). Use channel 2 to measure the peak-to-peak voltage across the load resistor (8.2 kΩ). Determine the voltage gain from the function generator to the final output. What is this voltage gain in decibels?

$A_V =$ _____. $A_{V(dB)} =$ _____.

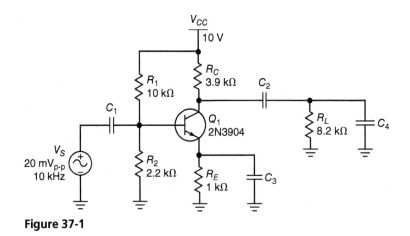

Figure 37-1

4. Vary the frequency of the input as needed to locate the lower cutoff frequency. Record its value here:

$f_1 =$ _____.

5. Measure and record the voltage gain one decade below the cutoff frequency recorded in Step 4.

$A_{V(dB)} =$ _____.

6. In this circuit, C_1 is producing the dominant cutoff frequency. Change C_1 from 0.1 μF to 0.47 μF. Using Eq. 14-30 from the textbook, calculate the new value of the lower cutoff frequency.

Answer = _____.

7. Repeat Step 4:

$f_1 =$ _____.

8. Vary the frequency of the input as needed to locate the upper cutoff frequency. Record its value here:

$f_2 =$ _____.

9. Change C_4 from 470 pF to 220 pF. Estimate the new value of the upper cutoff frequency.

Answer = _____.

10. Repeat Step 8:

$f_2 =$ _____.

11. The wiring of the circuit may have some effect on the upper cutoff frequency. Explain why this is so. Use the space on the experiment sheet.

12. If a spreadsheet is available, plot the voltage gain (y-axis) versus frequency (x-axis) using a scatter graph. If a spreadsheet is not available, sketch the ideal Bode plot and label the important parts of the graph.

13. Restore the original circuit of Fig. 37-1 by making $C_1 = 0.1$ μF and $C_4 = 470$ pF. This time, use a square-wave input signal with a peak-to-peak value of 20 mV and a frequency of 10 kHz. Look at the output signal. Sketch the waveform on the experiment sheet.

14. The normally horizontal lines of the square wave are now sloped at an angle. This effect is called *sag*. The charging and discharging of the dominant coupling capacitor C_1 are causing this sag. Change C_1 from 0.1 μF to 0.47 μF and notice how the sag decreases.

15. Increase the frequency of the input square wave to 50 kHz. Measure the risetime and record the value here:

$T_R =$ _____.

16. Calculate the upper cutoff frequency and record its value here:

$f_2 =$ _____.
This value should be in reasonable agreement with the value recorded in Step 8.

17. Change C_4 from 470 pF to 220 pF. Estimate the new risetime.

$T_R =$ _____.

18. Repeat Step 15:

$T_R =$ _____.

19. Remove C_4. Measure and record the risetime.

$T_R =$ _____.

Experiment 37

Lab Partner(s) _____

PARTS USED Nominal Value	Measured Value	PARTS USED Nominal Value	Measured Value	PARTS USED Nominal Value	Measured Value
1 kΩ		10 kΩ		0.47 μF	
2.2 kΩ		220 pF		10 μF	
3.9 kΩ		470 pF		470 μF	
8.2 kΩ		0.1 μF			

Explain the effects of the wiring on the upper cutoff frequency (f_2):

Sketch the waveform due to the square wave input here:

Questions for Experiment 37

1. If C_1 is dominant, an increase in C_1 has what effect on the lower cutoff frequency?   ( )
   (a) decreases it;   (b) doubles it;   (c) increases it;   (d) no effect.
2. What effect does a decrease in C_4 have on the upper cutoff frequency?   ( )
   (a) decreases it;   (b) doubles it;   (c) increases it;   (d) no effect.
3. If C_1 is dominant, an increase in C_1 has what effect on the sag?   ( )
   (a) decreases it;   (b) doubles it;   (c) increases it;   (d) no effect.
4. What effect does a decrease in C_4 have on the risetime?   ( )
   (a) decreases it;   (b) doubles it;   (c) increases it;   (d) no effect.
5. In this experiment, the decibel voltage gain was closest to: ( )
   (a) 0 dB;   (b) 20 dB;   (c) 40 dB;   (d) 60 dB.
6. To get minimum risetime, the stray-wiring capacitance across the load   ( )
   must be:
   (a) minimum;   (b) maximum;   (c) equal to the output coupling
   capacitor;   (d) none of these.
7. To measure risetime, the input signal must be a: ( )
   (a) sine wave;   (b) square wave;   (c) triangular wave;   (d) none of these.

8. If the voltage gain of an amplifier is 10, the voltage gain at the upper cutoff frequency is: ( )
   (a) −3 dB;   (b) 7 dB;   (c) 13 dB;   (d) 7.07.
9. Optional. Instructor's question.

10. Optional. Instructor's question.

Frequency Response of Coupling and Bypass Capacitors

Capacitors are used to couple signal sources to an amplifier and multiple stages together. In addition, capacitors are used to bypass an ac signal to ground for a variety of reasons. Because capacitors are widely used in electronic systems, it is very important that their response to the frequency of the signal passing through them is understood. As an example, the amplifiers used in previous experiments were operated at a frequency within the amplifier's midband. While operating at frequency within their midband, the coupling capacitors are transparent to the signal passing through them. However, if the signal passing through the amplifier is at a frequency outside of the amplifier's midband, the capacitors are no longer transparent to the signal passing through them. Outside of the midband, some of the signal is dropped across the coupling capacitors. The same is true for the bypass capacitors.

In this experiment, the frequency response of both coupling and bypass capacitors will be explored.

GOOD TO KNOW

The Bode plot is a graphical representation of the frequency response of a circuit. Spreadsheets can be used to graph the measured data to produce the Bode plot. Due to the wide range of frequencies, the x-axis is usually expressed in logarithmic scale.

Required Reading

Chapter 14 of *Electronic Principles,* 8th ed.

Equipment

1   function generator
3   resistors: 1 kΩ, 3.9 kΩ, 8.2 kΩ
3   capacitors: 0.1 μF, 2.2 μF, 10 μF
1   DMM (digital multimeter)
1   oscilloscope

Procedure

FREQUENCY RESPONSE

1. The RC combination in Fig. SA4-1 forms a voltage divider. As the frequency of the input signal varies, so does the capacitive reactance of C. As the frequency of V_S increases, the value of X_C decreases, as does the voltage across C.

2. Using a DMM, measure and record the value of the resistors and capacitors. Build the circuit in Fig. SA4-1. Set

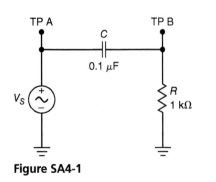

Figure SA4-1

$$X_c = \frac{1}{2\pi f C}$$

$$X_c = \frac{1}{2\pi (10\ Hz)(0.1\ \mu F)}$$

$$X_c = 159.154\ k\Omega$$

the output of the function generator for 1 V_P at a 10-Hz sine wave. Measure the voltage at test point B with the oscilloscope. Record the measured value in Table SA4-1. (*Note:* It is a good idea to leave channel 1 of the oscilloscope connected to test point A so the input signal can also be viewed.)

3. The cutoff (or corner) frequency, where the output voltage is 0.707 times the midband voltage, can be determined by using the formula $f_1 = \frac{1}{2\pi RC}$

4. Calculate the cutoff frequency for the circuit shown in Fig. SA4-1. Record the calculated value in Table SA4-2.

5. Set the input signal V_S to the remaining values listed in Table SA4-1. Record the amplitude of the signal at test point B for each input signal value in Table SA4-1.

6. Draw the ac model for the circuit in Fig. SA4-2 in the space provided on the data sheet. Redraw the ac model using V_S, C_1, and $z_{instage}$ as a single resistance. Compare the drawing to Fig. SA4-1.

   The output side of the ac model includes a current source and a parallel resistance. Using the techniques used to convert a Thevenin to a Norton circuit, the current source and parallel resistance can be replaced with a voltage source and series resistance, as shown in Fig. SA4-3.

7. Calculate the cutoff frequency for the Thevenin-equivalent circuit in Fig. SA4-3. The sum of R_C and R_L should be used for "R" in the given formula:
$$f_1 = \frac{1}{2\pi RC}$$

8. Record the calculated value in Table SA4-2.

9. The bypass capacitor C_3 forms a voltage divider with z_{out}, the Thevenin resistance of the amplifier in Fig. SA4-2 facing it, as shown in Fig. SA4-4. The formula to calculate z_{out} is

$$z_{out} = R_E \| \left(r_e' + \frac{R_G \| R_1 \| R_2}{\beta} \right)$$ (*Note:* Use $\beta = 100$ and no R_G is used.)

10. Calculate the cutoff frequency for the circuit in Fig. SA4-4. The z_{out} should be used for "R" in the given formula: $f_1 = \frac{1}{2\pi RC}$

11. Record the calculated value in Table SA4-2.

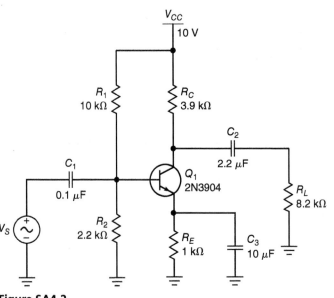

Figure SA4-2

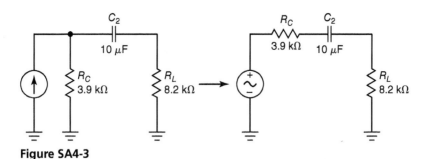

Figure SA4-3

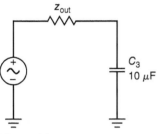

Figure SA4-4

TROUBLESHOOTING

12. If the wrong value capacitor was accidently used for either the coupling capacitor or the bypass capacitor, their respective cutoff frequencies would change

values. If a 10 μF capacitor was used instead of a 0.1 μF for C_1, how would it affect the cutoff frequency for Fig. SA4-1? (Be specific.)

CRITICAL THINKING

13. Optional: Multisim has a virtual Bode plotter. This virtual device will sweep the input frequency of the voltage source while measuring the output voltage. The results are displayed in graphical form. An example is shown in Fig. SA4-5.

14. Build the other two circuits, SA4-3 and SA4-4, in Multisim. Using the virtual Bode plotter, determine the corner frequency for the two circuits.

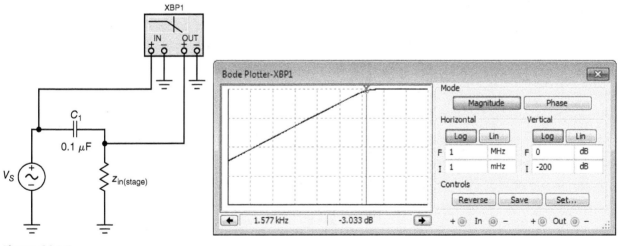

Figure SA4-5

System Application 4

Lab Partner(s)_____

PARTS USED Nominal Value	Measured Value
1 kΩ	
3.9 kΩ	
10 kΩ	
0.1 μF	
2.2 μF	
10 μF	

TABLE SA4-1 OUTPUT SIGNAL MEASUREMENTS

V_S, 1-V_P sine wave at	Multisim	Actual
10 Hz		
100 Hz		
500 Hz		
1 kHz		
1.2 kHz		
1.3 kHz		
1.4 kHz		
1.5 kHz		
1550 Hz		
1585 Hz		
1600 Hz		
1.7 kHz		
5 kHz		
10 kHz		
100 kHz		

AC Model of Figure SA4-2

AC model redrawn with just V_S, C_1, and $z_{instage}$ as a single resistance.

TABLE SA4-2 CUTOFF FREQUENCY

Cutoff Frequency	Calculated
Figure SA4-1	
Figure SA4-3	
Figure SA4-4	

Questions for System Application 4

1. The output voltage drops to _____ of the midband voltage at the cutoff frequency.
   (a) 0.707;   (b) 1.414;   (c) 0.63;   (d) 0.5.
2. Coupling capacitors should _____ affect the signal passing through them.
   (a) minimally;   (b) largely;   (c) not;   (d) always.
3. A _____ is a graphical representation of frequency response.
   (a) time constant;   (b) dielectric curve;   (c) knee voltage;   (d) Bode plot.
4. The coupling capacitor should look like a(n) _____ to an ac signal.
   (a) nonlinear device;   (b) voltage divider;   (c) short;   (d) open.
5. The bypass capacitor should look like a(n) _____ to an ac signal.
   (a) nonlinear device;   (b) voltage divider;   (c) short;   (d) open.
6. The output signal drops _____ from the midband value at the cutoff frequency.
   (a) 3 dB;   (b) 6 dB;   (c) 20 dB;   (d) 0.707 dB.
7. Another name for the corner frequency is the _____ frequency.
   (a) second;   (b) resonant;   (c) center;   (d) cutoff.
8. The x-axis of a Bode plot represents _____.
   (a) dB;   (b) frequency;   (c) phase shift;   (d) farads.
9. The y-axis of a Bode plot represents _____.
   (a) dB;   (b) frequency;   (c) phase shift;   (d) farads.
10. The x-axis of a Bode plot is usually expressed in _____ scale.
    (a) binary;   (b) metric;   (c) logarithmic;   (d) absolute.

The Differential Amplifier

The differential amplifier is the direct-coupled input stage of the typical op amp. The most common form of a diff amp is the double-ended input and single-ended output circuit. Some of the important characteristics of a diff amp are the input offset current, input bias current, input offset voltage, and common-mode rejection ratio. In this experiment, a diff amp will be built and the foregoing quantities measured.

GOOD TO KNOW

Voltage gain is often expressed in decibels. The formula to convert A_V to $A_{V(dB)}$ is:

$$A_{V(dB)} = 20 \log (A_V).$$

Required Reading

Chapter 15 (Secs. 15-1 to 15-5) of *Electronic Principles,* 8th ed.

Equipment

1 signal generator
2 power supplies: ± 15 V
10 ½-W resistors: two 22 Ω, two 100 Ω, two 1.5 kΩ, two 4.7 kΩ, two 10 kΩ (5% tolerance)
2 transistors: 2N3904
1 capacitor: 0.47 μF
1 DMM (digital multimeter)
1 oscilloscope

Procedure

TAIL CURRENT AND BASE CURRENTS

1. Measure and record the resistor and capacitor values. Notice the pair of swamping resistors (22 Ω) in Fig. 38-1. These have to be included in this experiment to improve the match between the discrete transistors. In Fig. 38-1, assume the typical h_{FE} is 200. Calculate the approximate tail current. Record in

Table 38-1. Also calculate and record the base current in each transistor.
2. Build the circuit shown in Fig. 38-1.
3. Measure and record the tail current.
4. Use the DMM as an ammeter to measure the base current in each transistor. If the DMM is not sensitive enough to measure microampere currents, then use the oscilloscope on dc input to measure the voltage across each base resistor and calculate the base current. Record the base currents in Table 38-1.

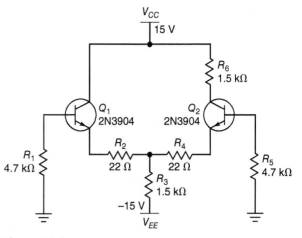

Figure 38-1

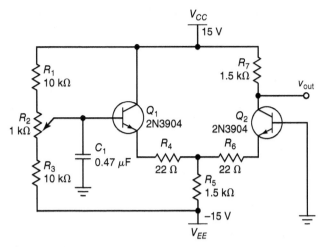

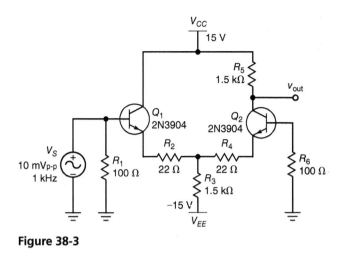

Figure 38-2

Figure 38-3

INPUT OFFSET AND BIAS CURRENTS

5. With the calculated data of Table 38-1, calculate the values of input offset current and input bias current. Record the theoretical answers in Table 38-2.
6. With the measured data of Table 38-1, calculate the values of $I_{in(off)}$ and $I_{in(bias)}$. Record the experimental answers in Table 38-2.

OUTPUT OFFSET VOLTAGE

7. In Fig. 38-2, assume that the base of Q_1 is grounded by a jumper wire. If both transistors are identical and all components have the values shown, then the dc output voltage would have a value of approximately +7.85 V. For this part of the experiment, any deviation from +7.85 V is called *output offset voltage*, designated $V_{out(off)}$.
8. Build the circuit shown in Fig. 38-2. Ground the base of Q_1 with a jumper wire. Measure the dc output voltage. Calculate the output offset voltage and record $V_{out(off)}$ in Table 38-3.
9. Remove the ground from the Q_1 base. Adjust the potentiometer until the output voltage is +7.85 V.
10. Measure the base voltage of Q_1. Record in Table 38-3 as $V_{in(off)}$.

DIFFERENTIAL VOLTAGE GAIN

11. Because of the swamping resistors in Fig. 38-3, the differential voltage gain is given by $R_C/2(r_E + r'_e)$. Calculate and record A_V in Table 38-4.
12. Build the circuit. Set the signal generator at 1 kHz with a signal level of 10 mV$_{p-p}$.
13. Measure the output voltage. Calculate and record the experimental value of A_V.

COMMON-MODE VOLTAGE GAIN

14. Calculate the common-mode voltage gain of Fig. 38-3. Record $A_{V(CM)}$ in Table 38-4.
15. Put a jumper wire between the bases of the circuit.
16. Increase the signal level until the output voltage is 0.5 V$_{p-p}$.
17. Measure the peak-to-peak input voltage. Calculate and record the experimental value of $A_{V(CM)}$.

COMMON-MODE REJECTION RATIO

18. Calculate and record the theoretical value of CMRR using the calculated data of Table 38-4.
19. Calculate and record the experimental value of CMRR using the experimental data of Table 38-4.

TROUBLESHOOTING

20. In this part of the experiment, a collector-emitter short means that all three transistor terminals are shorted together. A collector-emitter open means the transistor is removed from the circuit.
21. In Fig. 38-3, estimate the dc output voltage for each trouble listed in Table 38-5.
22. Insert each trouble; measure and record the dc voltages of Table 38-5.

CRITICAL THINKING

23. Select resistance values in Fig. 38-3 to get a tail current of 3 mA and a dc output voltage of +7.5 V. Record the nearest standard values in Table 38-6.
24. Build the circuit with the design values. Measure and record the tail current and dc output voltage.

Experiment 38

Lab Partner(s) _____

PARTS USED Nominal Value	Measured Value	PARTS USED Nominal Value	Measured Value	PARTS USED Nominal Value	Measured Value
22 Ω		1.5 kΩ		10 kΩ	
22 Ω		1.5 kΩ		10 kΩ	
100 Ω		4.7 kΩ		0.47 μF	
100 Ω		4.7 kΩ			

TABLE 38-1. TAIL AND BASE CURRENTS

	Calculated	Measured	
		Multisim	Actual
I_T			
I_{B1}			
I_{B2}			

TABLE 38-2. INPUT OFFSET AND BIAS CURRENTS

	Theoretical	Experimental	
		Multisim	Actual
$I_{in(off)}$			
$I_{in(bias)}$			

TABLE 38-3. INPUT AND OUTPUT OFFSET VOLTAGES

	Multisim	Actual
$V_{out(off)}$		
$V_{in(off)}$		

TABLE 38-4. VOLTAGE GAINS AND CMRR

	Calculated	Measured	
		Multisim	Actual
A_V			
$A_{V(CM)}$			
CMRR			

TABLE 38-5. TROUBLESHOOTING

Trouble	Estimated V_{out}	Measured V_{out}	
		Multisim	Actual
Q_1 C-E shorted			
Q_1 C-E open			
Q_2 C-E shorted			
Q_2 C-E open			

TABLE 38-6. CRITICAL THINKING: $R_E =$ _____; $R_C =$ _____

Multisim	% Error	Actual	% Error

Questions for Experiment 38

1. The tail current of Table 38-1 is closest to: ( )
   (a) 1 μA; (b) 23.8 μA; (c) 47.6 μA; (d) 9.53 mA.
2. The calculated base current of Fig. 38-1 is approximately: ( )
   (a) 1 μA; (b) 23.8 μA; (c) 47.6 μA; (d) 9.53 mA.
3. The calculated input bias current of Fig. 38-1 is approximately: ( )
   (a) 1 μA; (b) 23.8 μA; (c) 47.6 μA; (d) 9.53 mA.
4. The input offset voltage is the input voltage that removes the: ( )
   (a) tail current; (b) dc output voltage; (c) output offset voltage;
   (d) supply voltage.
5. The CMRR of Table 38-4 is closest to: ( )
   (a) 0.5; (b) 27.5; (c) 55; (d) 123.
6. Why is a high CMRR an advantage with a diff amp?

TROUBLESHOOTING

7. In Fig. 38-3, 150 Ω instead of 1.5 kΩ is used for the tail resistor. What are some of the dc and ac symptoms a technician can expect?

8. While troubleshooting the circuit of Fig. 38-3, what ac voltage should an oscilloscope display at the junction of the 22 Ω resistor with respect to ground?

CRITICAL THINKING

9. What value was used for R_E and R_C in the design? What is the new value of CMRR?

10. Optional. Instructor's question.

212

Differential-Amplifier Supplement

The preceding experiment on the differential amplifier focused on a single-ended output. In this experiment, the double-ended output will be the focus. Recall the basic theory given in the textbook. A double-ended output has twice as much voltage gain as a single-ended output. Also, the outputs from the collectors have equal magnitudes but opposite phases.

GOOD TO KNOW

When using a variable resistor, it is common practice to include a fixed-value series resistor to limit the minimum resistance and maximum current flowing through it.

Required Reading

Chapter 15 (Secs. 15-1 and 15-5) of *Electronic Principles*, 8th ed.

Equipment

1 signal generator
2 power supplies: ±10 V
1 DMM (digital multimeter)
3 transistors: 2N3904
9 ½-W resistors: two 1 kΩ, one 3.9 kΩ, three 4.7 kΩ, three 10 kΩ
3 47-µF capacitors
1 potentiometer: 10 kΩ
1 oscilloscope

Procedure

SINGLE-ENDED INPUT AND DOUBLE-ENDED OUTPUT

1. Measure and record the resistor and capacitor values. Build the circuit shown in Fig. 39-1.
2. Reduce the generator amplitude to zero. Then, measure the dc collector voltage of Q_2. Vary the emitter resistance until you have approximately 5 V on the Q_2 collector.
3. Measure and record the dc voltages listed in Table 39-1.

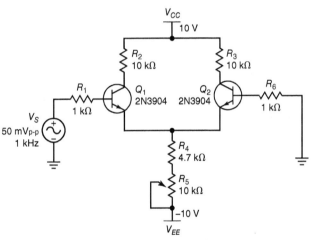

Figure 39-1

4. Use channel 1 of the oscilloscope to look at the ac base voltage of Q_1. Adjust the signal generator to get a frequency of 1 kHz and a sinusoidal amplitude of 50 mV$_{p-p}$.
5. Use channel 2 to look at the ac collector voltage of Q_1. An amplified and inverted sine wave should be present. Record the peak-to-peak value and phase in Table 39-2. (*Note:* The phase is recorded as an example in this first measurement.)
6. Repeat Step 5 for Q_2 and the top of the tail resistor (where the two emitters connect).

213

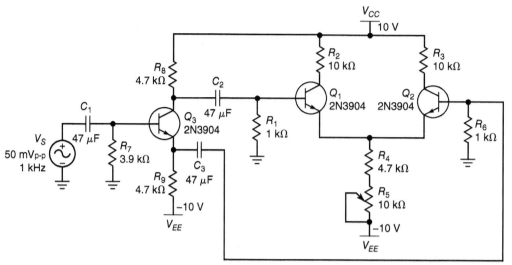

Figure 39-2

7. Turn off the power and reconnect the circuit with the signal generator on the other side of the diff amp.
8. Repeat Steps 5 and 6. (*Note:* Use Table 39-3.)

DIFFERENTIAL MODE

9. Build the circuit shown in Fig. 39-2. The first stage is called a *phase splitter,* a circuit that ideally produces equal-magnitude and opposite-phase signals.
10. Turn on the power. Repeat Step 2 to check that the dc collector voltage on Q_2 is still approximately 5 V.
11. Adjust the signal generator to get 1 kHz and a sinusoidal amplitude of 50 mV$_{p-p}$ at the base of Q_1. Table 39-4 shows the phase of v_1 as 0° because this signal will be used as a reference for other measurements.
12. Look at the ac collector voltage of Q_1. An amplified and inverted sine wave should be present. Record the peak-to-peak value and phase in Table 39-4.
13. Repeat Step 12 for the other voltages listed in Table 39-4. *Note:* When measuring v_{out}, use the oscilloscope in the

difference mode with channel 1 measuring v_{C2} and channel 2 measuring v_{C1}. If the oscilloscope does not include the difference function, mentally calculate the difference between v_{C2} and v_{C1}.

14. Use the data of Table 39-4 to calculate the voltage gain of Q_1 and Q_2:

Gain of $Q_1 =$ _____.

Gain of $Q_2 =$ _____.

COMMON-MODE OPERATION

15. Turn off the power and build the circuit shown in Fig. 39-3.
16. Apply power. Measure and record all quantities listed in Table 39-5.
17. Use the data of Table 39-5 to calculate the voltage gain of Q_1 and Q_2:

Gain of $Q_1 =$ _____.

Gain of $Q_2 =$ _____.

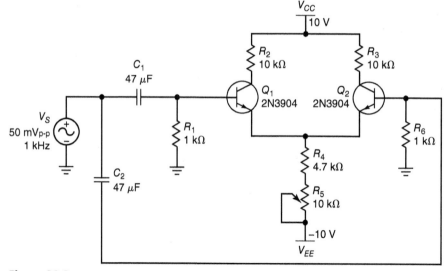

Figure 39-3

Experiment 39

Lab Partner(s) _____

PARTS USED Nominal Value	Measured Value	PARTS USED Nominal Value	Measured Value	PARTS USED Nominal Value	Measured Value
1 kΩ		4.7 kΩ		10 kΩ	
1 kΩ		4.7 kΩ		47 μF	
3.9 kΩ		10 kΩ		47 μF	
4.7 kΩ		10 kΩ		47 μF	

TABLE 39-1. DC VOLTAGES

	V_{C1}	V_{C2}	V_E	V_{EE}	V_{B1}	V_{B2}
Multisim						
Actual						

TABLE 39-2. NONINVERTING INPUT

	Multisim				Actual			
	v_1	v_{c1}	v_{c2}	v_e	v_1	v_{c1}	v_{c2}	v_e
Magnitude								
Phase								

TABLE 39-3. INVERTING INPUT

	Multisim				Actual			
	v_2	v_{c1}	v_{c2}	v_e	v_2	v_{c1}	v_{c2}	v_e
Magnitude	50 mV							
Phase	0°							

TABLE 39-4. DIFFERENTIAL INPUT

	Multisim		Actual	
	Magnitude	Phase	Magnitude	Phase
v_1	50 mV	0°	50 mV	0°
v_{c1}				
v_2				
v_{c2}				
v_e				
v_{out}				

TABLE 39-5. COMMON-MODE INPUT

	Multisim		Actual	
	Magnitude	Phase	Magnitude	Phase
v_1	50 mV	0°	50 mV	0°
v_{c1}				
v_2				
v_{c2}				
v_e				
v_{out}				

Questions for Experiment 39

1. In Table 39-2, compare and explain the phase of v_{c1} and v_{c2} with the reference phase of v_1:

2. In Table 39-3, compare and explain the phase of v_{c1} and v_{c2} with the reference phase of v_2:

3. Using the data of Table 39-2, calculate the single-ended voltage gain of each transistor. Also, calculate the differential voltage gain. Be sure to show the calculations:

4. Using the data of Table 39-4, explain the significance of the waveforms at v_{c1} and v_{c2}.

5. Using the data of Table 39-4, explain the significance of the waveform at v_e.

6. Compare the waveforms of Table 39-4 and Table 39-5. Now, explain why differential voltage gain is much greater than common-mode voltage gain.

Introduction to Op-Amp Circuits

When connecting op-amp circuits, the following precautions need to be taken to avoid possible damage or other unwanted effects:

1. Adjust the supply voltages to the desired level. Then, turn the power off and connect the power supplies to the op-amp pins.

2. To avoid damaging an op amp, the power should remain off while breadboarding the rest of the circuit.

3. Make sure that the ac source is turned off before connecting it to the op-amp circuit.

4. To avoid oscillations and other unwanted effects, make sure that all wiring is as short as possible.

5. After a final wiring check, the dc supply voltages can be turned on.

6. Next, the ac source can be turned on. To avoid possible damage to the op amp, the ac peak voltages should always be less than the supply voltages.

7. If oscillations appear, use decoupling capacitors on the supply pins. Manufacturers recommend capacitances between 0.1 and 1 μF between each supply pin and ground.

8. When powering down, reduce the ac source to zero. Be sure to do this before turning off the dc supplies.

9. The last step is to turn off the dc supply voltages.

GOOD TO KNOW

Pin 1 is located to the left of the notch on the 741C IC.

Offset null	1		8	NC
Inverting input	2 −		7	V+
Noninverting input	3 +		6	Output
V−	4		5	Offset null

Required Reading

Chapter 16 (Secs. 16-1 and 16-2) of *Electronic Principles,* 8th ed.

Equipment

1 function generator
2 power supplies: ±15 V
1 op amp: 741C
5 ½-W resistors: two 100 Ω, 10 kΩ, two 100 kΩ
2 capacitors: 0.47 μF

2 potentiometers: 10 kΩ
1 DMM (digital multimeter)
1 oscilloscope

Procedure

1. Measure and record the resistor and capacitor values. Build the circuit shown in Fig. 40-1.

2. Use channel 1 of the oscilloscope to measure V_1, the dc voltage between the noninverting input (pin 3) and ground. *Note:* Use a sensitive range because the dc voltage will be less than ±15 mV.

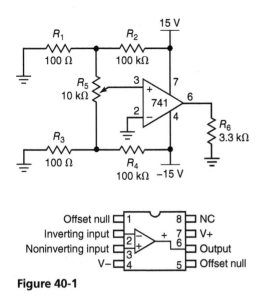

Figure 40-1

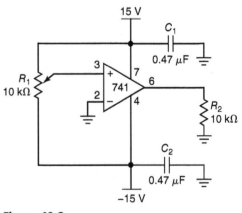

Figure 40-2

3. Use channel 2 of the oscilloscope to measure the dc output voltage (pin 6).

4. Vary the potentiometer. Somewhere near the middle of the range, the dc output voltage will be observed to change from positive to negative, or vice versa.

5. Slowly vary the potentiometer until the output goes into positive saturation. Record V_1 and V_{out} in Table 40-1.

6. Slowly vary the potentiometer until the output goes into negative saturation. Record V_1 and V_{out} in Table 40-1.

7. Repeat Steps 1 through 6 for Fig. 40-2. Record the data in Table 40-2.

8. The two circuits (Figs. 40-1 and 40-2) behave differently. What was observed, and how can the difference be explained?

9. In Fig. 40-3, potentiometer R_2 can be adjusted to produce the V_2 voltages shown in Table 40-3. For each V_2 voltage, what is the V_1 voltage that causes the output to switch states? Record the answers under "Calculated V_1."

10. Build the circuit shown in Fig. 40-3. Adjust R_2 to get each V_2 voltage listed in Table 40-3. Vary R_1 until the output switches states. Record the V_1 values in Table 40-3.

11. Repeat Steps 9 and 10 for Fig. 40-4. Record the answers in Table 40-4.

12. Build the circuit shown in Fig. 40-5 using the triangular output of a function generator.

13. Use channel 1 to look at the triangular input voltage.

14. Use channel 2 to look at the output voltage.

15. Adjust the amplitude of the triangular wave to 8 V peak to peak and the frequency to 100 Hz.

16. Vary the potentiometer over a wide range and notice how the output duty cycle changes.

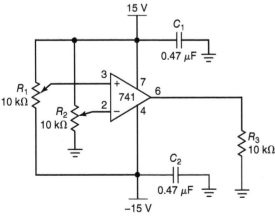

Figure 40-3

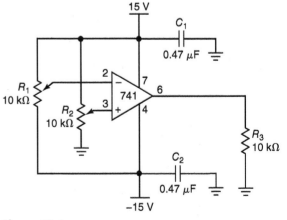

Figure 40-4

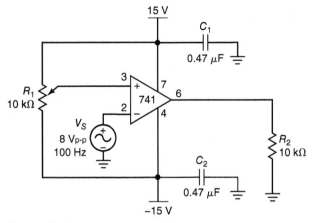

Figure 40-5

17. Explain the results of Step 16. Why did the duty cycle change?

18. Reverse the two input signals by connecting the triangular input to pin 3 and the potentiometer to pin 2.
19. Repeat Steps 15 and 16.
20. Discuss the phase relations of Step 19.

Experiment 40

Lab Partner(s) _____

PARTS USED Nominal Value	Measured Value	PARTS USED Nominal Value	Measured Value	PARTS USED Nominal Value	Measured Value
100 Ω		100 kΩ		0.47 μF	
100 Ω		100 kΩ		0.47 μF	
10 kΩ					

TABLE 40-1. FINE-ADJUSTMENT CIRCUIT

	Multisim		Actual	
	(+) Saturation	(−) Saturation	(+) Saturation	(−) Saturation
V_1				
V_{out}				

TABLE 40-2. COURSE-ADJUSTMENT CIRCUIT

	Multisim		Actual	
	(+) Saturation	(−) Saturation	(+) Saturation	(−) Saturation
V_1				
V_{out}				

TABLE 40-3. NONINVERTING COMPARATOR

	Calculated	Measured V_1	
V_2	V_1	Multisim	Actual
0			
1			
2			
5			
10			

TABLE 40-4. INVERTING COMPARATOR

V_2	Calculated V_1	Measured V_1 Multisim	Measured V_1 Actual
0			
1			
2			
5			
10			

Questions for Experiment 40

1. In Fig. 40-1, the maximum positive voltage to pin 3 is: ( )
   (a) 0;   (b) 15 mV;   (c) 30 mV;   (d) 15 V.
2. In Fig. 40-2, the wiper of the potentiometer is much closer to the upper end. The ( )
   output of the op amp is closest to:
   (a) 0;   (b) 15 mV;   (c) 30 mV;   (d) 15 V.
3. In Fig. 40-3, R_2 is adjusted to get $V_2 = +7.5$ V. To switch the output state, the ( )
   resistance between the wiper and the upper end of the potentiometer must be
   approximately:
   (a) 0;   (b) 1 kΩ;   (c) 2.5 kΩ;   (d) 5 kΩ.
4. Decoupling capacitors are used on the supply pins of the op amp to prevent: ( )
   (a) output switching;   (b) oscillations;   (c) damaging the op amp;
   (d) saturation.
5. The data of Table 40-1 demonstrates that the voltage gain of an op amp is very: ( )
   (a) low;   (b) high;   (c) sharp;   (d) unreliable.
6. The data of Table 40-3 demonstrate that the output switches when the noninverting ( )
   input voltage and the inverting input voltage are approximately:
   (a) equal;   (b) 180° out of phase;   (c) dc voltages;   (d) ac voltages.
7. Explain why the duty cycle changes in a circuit like Fig. 40-5.

8. What are the three most important things demonstrated in this experiment?

9. Optional. Instructor's question.

10. Optional. Instructor's question.

Inverting and Noninverting Amplifiers

Inverting and noninverting amplifiers are the most fundamental op-amp circuits. The closed-loop voltage gain of an inverting amplifier equals the ratio of the feedback resistance to the input resistance, $-R_f/R_1$. On the other hand, the closed-loop voltage gain of a noninverting amplifier equals the ratio of the feedback resistance to the input resistance plus 1, $R_f/R_1 + 1$.

In this experiment, both types of amplifiers will be built and their operation examined. In addition, their bandwidth will be measured and the slew-rate distortion observed. Finally, the effect of a decrease in load resistance on the MPP of the op amp will be investigated.

Required Reading

Chapter 16 (Secs. 16-1 to 16-4) of *Electronic Principles*, 8th ed.

Equipment

1   signal generator
2   power supplies: ±15 V
1   op amp: 741C
7   ½-W resistors: two 100 Ω, three 1 kΩ, two 10 kΩ
1   potentiometer: 1 kΩ
1   DMM (digital multimeter)
1   oscilloscope

Procedure

1. In Fig. 41-1, v_{in} is the voltage between the potentiometer wiper and ground. Calculate the output voltage in this circuit for each of the input voltages shown in Table 41-1. Record the values under v_{out} (calculated).

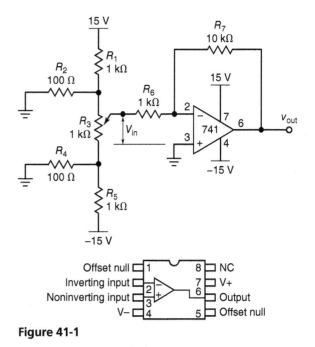

Figure 41-1

2. Measure and record the resistor values. Build the circuit shown in Fig. 41-1.

3. To keep the wiring simple in this experiment, the decoupling capacitors that are normally connected between the supply pins and ground have been omitted. If the output signal is very noisy or other unwanted effects appear, add 0.47 μF capacitors to pins 4 and 7, as shown in Experiment 40.

4. Use channel 1 of the oscilloscope to measure v_{in}, and use channel 2 to measure v_{out}. Vary the potentiometer over its full range while observing the input and output waveforms.

5. Adjust the potentiometer to produce each positive input voltage listed in Table 41-1. Measure and record the output voltage for each input voltage.

6. Table 41-2 shows negative input voltages for the circuit of Fig. 41-1. What are the theoretical output voltages? Record the calculated values in Table 41-2.

7. Measure and record the output voltage for each input voltage in Table 41-2.

8. Build the circuit shown in Fig. 41-2. Use channels 1 and 2 of the oscilloscope to look at the input and output voltages.

9. Adjust the input voltage to 2 V_{p-p} and 1 kHz. Notice how the input and output waveforms are related in magnitude and phase.

10. Repeat Step 9 for the following frequencies: 10 Hz, 100 Hz, and 10 kHz. What conclusions can be drawn about voltage gain and phase shift:

11. Repeat Step 9 for 100 kHz. Why does this happen?

12. Set the frequency to 1 kHz.

13. In Table 41-3, what is the peak-to-peak output voltage for each input voltage shown? Record the calculated results.

14. Measure the peak-to-peak output voltage for each input voltage of Table 41-3. Record the measured values.

15. Build the circuit shown in Fig. 41-3. Use channels 1 and 2 of the oscilloscope to view the input and output voltages.

16. Adjust the input voltage to 2 V_{p-p} and 1 kHz. Notice how the input and output waveforms are related in magnitude and phase.

17. In Table 41-4, what is the peak-to-peak output voltage for each input voltage shown? Record the calculated results.

18. Measure the peak-to-peak output voltage for each input voltage of Table 41-4. Record the measured values.

19. Adjust the generator amplitude to get an output voltage with a peak-to-peak value of 1 V. Increase the generator frequency until the output amplitude decreases to 0.7 V_{p-p}. Record the frequency as the bandwidth in Table 41-5. Is there any slew-rate distortion? Record *yes* or *no*.

20. Set the frequency to 1 kHz and adjust the generator amplitude to get an output voltage of 20 V_{p-p}. Increase the generator frequency until the output amplitude is 14 V_{p-p}. Record the frequency as the bandwidth. Is there any slew-rate distortion? Record in Table 41-5.

21. Reduce the generator amplitude to zero and adjust the frequency to 1 kHz.

22. Connect a load resistor of 10 kΩ between pin 6 and ground.

23. Increase the generator amplitude until clipping starts on the output signal. Measure and record the peak-to-peak output voltage as the MPP in Table 41-6.

24. Change the load resistance to 100 Ω. Vary the generator amplitude until clipping starts on the output signal. Record the MPP for this load resistance.

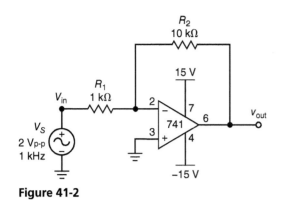

Figure 41-2

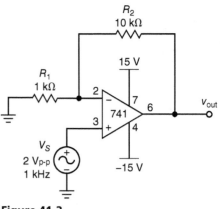

Figure 41-3

Experiment 41

Lab Partner(s) _____

PARTS USED Nominal Value	Measured Value	PARTS USED Nominal Value	Measured Value	PARTS USED Nominal Value	Measured Value
100 Ω		1 kΩ		10 kΩ	
100 Ω		1 kΩ			
1 kΩ		10 kΩ			

TABLE 41-1. INVERTING AMPLIFIER (POSITIVE INPUTS)

	Calculated	Measured v_{out}	
v_{in}	v_{out}	Multisim	Actual
0.1 V			
0.2 V			
0.5 V			
1 V			

TABLE 41-2. INVERTING AMPLIFIER (NEGATIVE INPUTS)

	Calculated	Measured v_{out}	
v_{in}	v_{out}	Multisim	Actual
−0.1 V			
−0.2 V			
−0.5 V			
−1 V			

TABLE 41-3. INVERTING AMPLIFIER (SINUSOIDAL INPUTS)

	Calculated	Measured v_{out}	
v_{in}	v_{out}	Multisim	Actual
0.1 V_{p-p}			
0.2 V_{p-p}			
0.5 V_{p-p}			
1 V_{p-p}			
2 V_{p-p}			

TABLE 41-4. NONINVERTING AMPLIFIER (SINUSOIDAL INPUTS)

v_{in}	Calculated v_{out}	Measured v_{out} Multisim	Actual
0.1 V_{p-p}			
0.2 V_{p-p}			
0.5 V_{p-p}			
1 V_{p-p}			
2 V_{p-p}			

TABLE 41-5. BANDWIDTH OF NONINVERTING AMPLIFIER

	Multisim BW	Slew-Rate Distortion	Actual BW	Slew-Rate Distortion
v_{out}				
1 V_{p-p}				
20 V_{p-p}				

TABLE 41-6. MAXIMUM PEAK-TO-PEAK OUTPUT

	MPP Multisim	MPP Actual
R_L		
10 kΩ		
100 Ω		

Questions for Experiment 41

1. The input voltage dividers of Fig. 41-1 reduce the total voltage across the potentiometer ( )
   to approximately:
   (a) 15 mV;   **(b)** 1.5 V;   **(c)** 3 V;   **(d)** 15 V.
2. In Fig. 41-1, the closed-loop voltage gain is: ( )
   (a) 1;   **(b)** 10;   **(c)** 11;   **(d)** 100.
3. In Fig. 41-2, the peak-to-peak output voltage at low frequencies is: ( )
   (a) 1 V;   **(b)** 2 V;   **(c)** 10 V;   **(d)** 20 V.
4. In Fig. 41-2, the voltage between pin 2 and ground is approximately: ( )
   (a) 0 V;   **(b)** 1 V;   **(c)** 2 V;   **(d)** 20 V.
5. The closed-loop voltage gain of Fig. 41-3 is approximately: ( )
   (a) 1;   **(b)** 10;   **(c)** 11;   **(d)** 100.
6. If the 10 kΩ resistor opens in Fig. 41-3, the output voltage is: ( )
   (a) 0;   **(b)** 10 V;   **(c)** sinusoidal;   **(d)** rectangular.
7. In Fig. 41-3, the product of the closed-loop voltage gain and the small-signal bandwidth ( )
   is closest to:
   (a) 1 kHz;   **(b)** 20 kHz;   **(c)** 90 kHz;   **(d)** 1 MHz.

8. Why did slew-rate distortion appear in Step 11 and not in Step 10?

9. Why was the MPP smaller in Step 24 than in Step 23?

10. Optional. Instructor's question.

42

The Operational Amplifier

An operational amplifier, or op amp, is a high-gain dc amplifier usable from 0 to over 1 MHz (typical). By connecting external resistors to an op amp, the voltage gain and bandwidth can be adjusted to meet the circuit requirements. A thorough understanding of the characteristics of the op amp will aid in both the design of a new circuit and the troubleshooting of an existing circuit. These include the input offset current, input bias current, input offset voltage, common-mode rejection ratio, MPP value, short-circuit output current, slew rate, and power bandwidth. In this experiment, a basic op-amp circuit will be built and analyzed.

GOOD TO KNOW

Sometimes the op amp's power supply connections are omitted from the schematic diagram. Be sure to always connect them, as the op amp is an active device.

Required Reading

Chapter 16 (Secs. 16-1 and 16-2) of *Electronic Principles*, 8th ed.

Equipment

1 signal generator
2 power supplies: ±15 V
8 ½-W resistors: two 100 Ω, 1 kΩ, two 10 kΩ, 100 kΩ, two 220 kΩ
3 op amps: 741C
2 capacitors: 0.47 μF
1 DMM (digital multimeter)
1 oscilloscope

Procedure

INPUT OFFSET AND BIAS CURRENTS

1. The 741C has a typical $I_{in(bias)}$ of 80 nA. Assume that this is the base current in each 220 kΩ resistor of Fig. 42-1. Calculate dc voltages at the noninverting and inverting inputs. Record in Table 42-1.

2. Measure and record the resistor and capacitor values. Build the circuit shown in Fig. 42-1.

3. Measure the dc voltage at the noninverting input. Record the measured data in Table 42-1.

4. Measure and record the inverting input voltage.

5. Repeat Steps 1 through 4 for the other 741Cs. If Multisim is being used, replace the LM741C op amp with LM725C and LM715C op amps for this step.

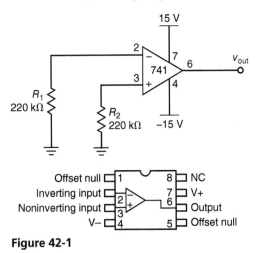

Figure 42-1

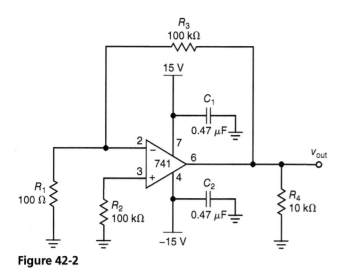

Figure 42-2

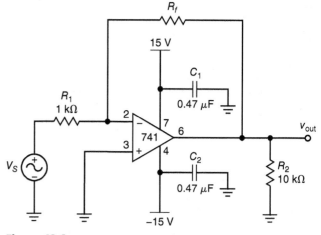

Figure 42-3

6. Using the measured data of Table 42-1, calculate the base currents, then the values of $I_{in(off)}$ and $I_{in(bias)}$. Record the results in Table 42-2.

OUTPUT OFFSET VOLTAGE

7. Build the circuit shown in Fig. 42-2. *Note:* Bypass capacitors are used on each supply voltage to prevent oscillations. These capacitors should be connected as close to the IC as possible.
8. Measure the dc output voltage. Record this value as $V_{out(off)}$ in Table 42-3.
9. Repeat Step 8 for the other 741Cs.
10. With the resistors shown in Fig. 42-2, the circuit has a voltage gain of 1000. Calculate the input offset voltage with

$$V_{in(off)} = \frac{V_{out(off)}}{1000}$$

Record the calculated values in Table 42-3.

MAXIMUM OUTPUT CURRENT

11. Disconnect the right end of the 100 kΩ resistor from the output.
12. Connect the right end of the 100 kΩ resistor to the +15 V. This will apply approximately 15 mV to the inverting input, more than enough to saturate the op amp.
13. Replace the 10 kΩ load resistor with a DMM used as an ammeter. Since the ammeter has a very low resistance, it indicates the short-circuit output current.
14. Read and record I_{max} in Table 42-3.
15. Repeat Step 14 for the other 741Cs.

SLEW RATE

16. Build the circuit shown in Fig. 42-3 with an R_f of 100 kΩ.
17. Use the oscilloscope (time base approx. 20 μs/DIV) to look at the output of the op amp. Set the signal generator at 5 kHz. Adjust the signal level to get hard

clipping on both peaks of the output signal (overdrive condition).
18. Measure the voltage change and the time change of the waveform. Calculate and record the slew rate in Table 42-4.
19. Repeat Step 18 for the other 741Cs.

POWER BANDWIDTH

20. Change R_f to 10 kΩ. Set the ac generator at 1 kHz. Adjust the signal level to get 20 V_{p-p} out of the op amp.
21. Increase the frequency from 1 to 20 kHz and watch the waveform. Somewhere above 8 kHz, slew-rate distortion will become evident because the waveform will appear triangular and the amplitude will decrease.
22. Record the approximate (ballpark) frequency where slew-rate distortion begins (Table 42-4).
23. Repeat Steps 20 to 22 for the other 741Cs.

MPP VALUE

24. Set the ac generator at 1 kHz. Increase the signal level until clipping just starts on either peak.
25. Record the MPP in Table 42-4.

TROUBLESHOOTING

26. Measure the dc and ac output voltage for each trouble listed in Table 42-5.
27. Record the measured values in Table 42-5.

CRITICAL THINKING

28. As derived in the textbook, the voltage gain of a circuit like Fig. 42-3 is equal to $-R_f/R_1$. Select a value of R_f to get a voltage gain of 68.
29. Replace R_f with the selected value. Measure the voltage gain.
30. Record the design value and the measured voltage gain in Table 42-6.

Experiment 42

Lab Partner(s) _____

PARTS USED Nominal Value	Measured Value	PARTS USED Nominal Value	Measured Value	PARTS USED Nominal Value	Measured Value
100 Ω		10 kΩ		0.47 μF	
100 Ω		100 kΩ		0.47 μF	
1 kΩ		220 kΩ			
10 kΩ		220 kΩ			

TABLE 42-1. INPUT VOLTAGES

Op Amp	Calculated		Measured			
			Multisim		Actual	
	v_1	v_2	v_1	v_2	v_1	v_2
1						
2						
3						

TABLE 42-2. INPUT OFFSET AND BIAS CURRENTS

Op Amp	Multisim		Actual	
	$I_{in(off)}$	$I_{in(bias)}$	$I_{in(off)}$	$I_{in(bias)}$
1				
2				
3				

TABLE 42-3. INPUT AND OUTPUT OFFSET VOLTAGES

Op Amp	Multisim			Actual		
	$V_{out(off)}$	$V_{in(off)}$	I_{max}	$V_{out(off)}$	$V_{in(off)}$	I_{max}
1						
2						
3						

TABLE 42-4. SLEW RATE, POWER BANDWIDTH, AND MPP VALUE

Op Amp	Multisim			Actual		
	S_R	f_{max}	MPP	S_R	f_{max}	MPP
1						
2						
3						

TABLE 42-5. TROUBLESHOOTING

	Multisim		Actual	
Trouble	V_{out} DC	V_{out} DC	V_{out} DC	V_{out} DC
No +15 V				
No −15 V				
Pin 2 shorted to GND				

TABLE 42-6. CRITICAL THINKING: $R_f =$ _____

Multisim		Actual	
A_V	% Error	A_V	% Error

Questions for Experiment 42

1. The calculated dc voltages in Table 42-1 are approximately: ( )
    (a) 1 mV;   **(b)** 5.6 mV;   **(c)** 12.3 mV;   **(d)** 17.6 mV.
2. The input bias current of Table 42-2 is closest to: ( )
    (a) 1 nA;   **(b)** 80 nA;   **(c)** 2 mA;   **(d)** 25 mA.
3. The short-circuit currents of Table 42-3 are closest to: ( )
    (a) 1 nA;   **(b)** 80 nA;   **(c)** 2 mA;   **(d)** 25 mA.
4. When the input frequency was much higher than the f_{max} of Table 42-4, the output ( )
    looked:
    (a) sinusoidal;   **(b)** triangular;   **(c)** square;   **(d)** undistorted. ( )
5. The MPP value of Table 42-4 is closest to:
    (a) 5 mV;   **(b)** 15 V;   **(c)** 25 V;   **(d)** 30 V.
6. Explain the meaning of the input offset current and input bias current.

TROUBLESHOOTING

7. Explain the meaning of the input offset voltage.

8. Describe how the slew rate was measured in this experiment.

CRITICAL THINKING

9. What value was used for R_f? Why?

10. Optional. Instructor's question.

Small- and Large-Signal Output Impedances

Although an input voltage can be applied as needed to null the output, this direct approach to nulling the output can produce unwanted drift or other degrading effects. This is why it is best to use the nulling method suggested on the data sheet. For a 741, the manufacturer recommends using a potentiometer between pins 1 and 5. In this experiment, a potentiometer will be used in this manner to null the output of a 741C op-amp circuit.

Data sheets list the open-loop output impedance $z_{out(OL)}$ of a 741C as 75 Ω. The closed-loop output impedance $z_{out(CL)}$ is usually much lower, typically less than 1 Ω. This typical value applies only to small signals.

When the signal is large, the output impedance increases. The large-signal output impedance is represented by the symbol $z_{out(MPP)}$. This is the output impedance of an op amp when it is producing the maximum peak-to-peak output signal. The value of $z_{out(MPP)}$ is equal to the load resistance that reduces the MPP to half of the unloaded value. Fig. 16-7b in the textbook shows that a load resistance of slightly more than 200 Ω reduces the MPP from 27 V (unloaded MPP) to 13.5 V. In this experiment, the approximate value of $z_{out(MPP)}$ will be measured.

GOOD TO KNOW

The $z_{out(MPP)}$ and the load resistance (R_L) create a voltage divider. When the voltage across the load resistance (R_L) is 1/2 of the unloaded voltage, $R_L = z_{out(MPP)}$.

Required Reading

Chapter 16 (Secs. 16-1 to 16-4) of *Electronic Principles*, 8th ed.

Equipment

1 signal generator
2 power supplies: ±15 V
1 op amp: 741C
8 ½-W resistors: 10 Ω, 100 Ω, 1 kΩ, three 10 kΩ, two 100 kΩ
1 capacitor: 4.7 μF

2 potentiometers: 1 kΩ, 10 kΩ
1 DMM (digital multimeter)
1 oscilloscope

Procedure

1. Assume $I_{in(bias)} = 80$ nA, $I_{in(off)} = 20$ nA, and $V_{in(off)} = 2$ mV in Fig. 43-1. Using Eqs. (16-7) through (16-10) in the textbook, calculate and record the error voltages listed in Table 43-1.

2. Measure and record the resistor and capacitor values. Build the circuit shown in Fig. 43-1. Measure and record the output error voltage V_{err} in Table 43-1.

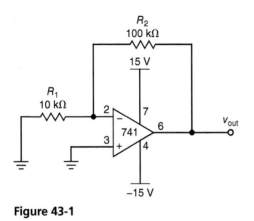

Figure 43-1

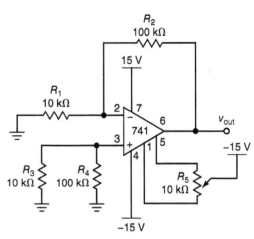

Figure 43-3

3. Note that the input error voltages are superimposed at the input terminals and cannot be measured separately. This is why they are listed as "DNA" (does not apply) in Table 43-1.

4. A circuit like Fig. 43-1 is not compensated for input bias current. When $I_{in(bias)}$ = 80 nA, V_{1err} ≈ 0.8 mV. This is for the typical value of input bias current for a 741C. In the worst case, $I_{in(bias)}$ = 500 nA and V_{1err} ≈ 5 mV. When V_{1err} is large like this, a designer may use a compensating resistor on the other side of the op amp to reduce the effect of input bias current.

5. Assume $I_{in(bias)}$ = 80 nA, $I_{in(off)}$ = 20 nA, and $V_{in(off)}$ = 2 mV in Fig. 43-2. Calculate and record the error voltages listed in Table 43-2.

6. Build the circuit shown in Fig. 43-2. Measure and record the output error voltage V_{err} in Table 43-2.

7. A circuit like Fig. 43-2 is compensated for the input bias current. Ideally, V_{1err} = 0, leaving only V_{2err} and V_{3err}. The best way to eliminate the effect of these remaining input error voltages is by using the nulling-potentiometer method shown in Fig. 43-3.

8. Build the circuit shown in Fig. 43-3.

9. With a DMM, adjust the potentiometer to get an output voltage as close to zero as possible.

10. Incidentally, the parallel equivalent resistance of 10 kΩ and 100 kΩ is 9.09 kΩ. Therefore, the nearest standard value of 9.1 kΩ can be used, instead of two separate resistors to compensate for input bias current.

11. After the circuit has been compensated and nulled, a signal can be applied to the circuit to measure small-signal and large-signal output impedances.

12. Connect a signal generator through a 4.7 μF coupling capacitor to the noninverting input, as shown in Fig. 43-4.

13. Set the generator frequency to 1 kHz.

14. Measure the output voltage with an oscilloscope. Adjust the generator amplitude to get an output sine wave with a peak-to-peak value of 1 V.

15. Connect a load resistance of 10 kΩ between pin 6 and ground. Measure and record the output voltage in Table 43-3.

16. Repeat Step 15 for the other load resistances shown in Table 43-3. Because of the heavy negative feedback, the closed-loop output impedance will be very small, and the load voltage should remain at approximately 1 V_{p-p} for all load resistances. In other words, the op amp should act like a stiff voltage source.

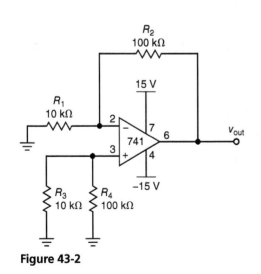

Figure 43-2

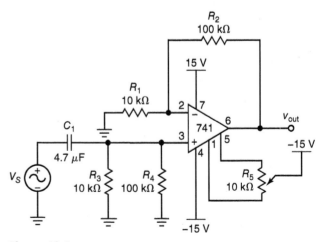

Figure 43-4

17. Disconnect the load resistor and increase the input signal until a maximum unclipped output voltage MPP is achieved. This should be 27 to 28 V_{p-p}. Make sure that the signal is not quite clipping by backing off slightly from the clipped level. Record the value here:

Unloaded MPP = _____.

18. Connect a load resistance of 10 kΩ between pin 6 and ground. If any clipping is visible on the oscilloscope, reduce the generator amplitude slightly to get the maximum unclipped output. Measure and record the peak-to-peak output voltage in Table 43-4.

19. Repeat Step 18 for the other load resistances shown in Table 43-4. After each load resistor is connected, the amplitude of the generator will need to be adjusted to prevent output clipping. Adjust the level as needed to get the maximum unclipped output voltage.

20. Notice how the MPP decreased in Step 19 as the load resistance decreased. The large-signal output impedance equals the load resistance when the loaded MPP is half of the unloaded MPP.

21. Replace the load resistor with a 1 kΩ potentiometer connected as a variable resistance. Vary the load resistance until the load voltage is half of the unloaded MPP. Disconnect the potentiometer and measure its resistance. Record its value here:

$z_{out(MPP)}$ = _____.

22. The foregoing value of output impedance applies to large signals only. The large-signal output impedance is always more than the small-signal open-loop output impedance z_{out} shown on data sheets.

Experiment 43

Lab Partner(s) _____

PARTS USED Nominal Value	Measured Value	PARTS USED Nominal Value	Measured Value	PARTS USED Nominal Value	Measured Value
10 Ω		10 kΩ		100 kΩ	
100 Ω		10 kΩ		100 kΩ	
1 kΩ		10 kΩ		4.7 μF	

TABLE 43-1. ERROR VOLTAGES WITHOUT COMPENSATION

	Calculated	Measured	
		Multisim	Actual
V_{1err}		DNA	DNA
V_{2err}		DNA	DNA
V_{3err}		DNA	DNA
V_{error}			

TABLE 43-2. ERROR VOLTAGES WITH COMPENSATION

	Calculated	Measured	
		Multisim	Actual
V_{1err}		DNA	DNA
V_{2err}		DNA	DNA
V_{3err}		DNA	DNA
V_{error}			

TABLE 43-3. SMALL-SIGNAL OUTPUT IMPEDANCE

R_L	10 kΩ	1 kΩ	100 kΩ	10 kΩ
V_L Multisim				
V_L Actual				

TABLE 43-4. LARGE-SIGNAL OUTPUT IMPEDANCE

R_L	10 kΩ	1 kΩ	100 kΩ	10 kΩ
V_L Multisim				
V_L Actual				

Questions for Experiment 43

1. In Fig. 43-1, the input error voltage caused by input bias current is typically:
   (a) 0.1;   (b) 0.8 mV;   (c) 2 mV;   (d) 6 mV.
2. The resistors connected to pin 3 of Fig. 43-2 compensate for the input: ( )
   (a) offset current;   (b) bias current;   (c) offset voltage;   (d) noise.
3. The 10-kΩ potentiometer of Fig. 43-3 is used to: ( )
   (a) set $A_{V(CL)}$;   (b) zero the output;   (c) improve ac response;   (d) add
   feedback.   ( )
4. The 4.7 μF capacitor prevents dc current from flowing through the:
   (a) generator resistance;   (b) 910 Ω;   (c) op-amp noninverting input;
   (d) 1 kΩ.   ( )
5. Connecting a small load resistance:
   (a) reduces MPP;   (b) increases MPP;   (c) increases gain;   (d) has no
   effect.   ( )
6. The output impedance of an op amp increases when the output signal is:
   (a) small;   (b) a dc signal;   (c) large;   (d) zero.   ( )
7. The closed-loop voltage gain of Fig. 43-4 is closest to: ( )
   (a) 1;   (b) 10;   (c) 11;   (d) 100.
8. The data of Table 43-3 implies that the small-signal output impedance is very: ( )
   (a) small;   (b) distorted;   (c) large;   (d) uncompensated.
9. Write a brief explanation of why Step 21 influenced the value of the large-signal output
   impedance.

10. Optional. Instructor's question.

Summing Amplifiers

The summing amplifier is an inverting amplifier with two or more input channels. Each channel has its own voltage gain given by the ratio of the feedback resistance to the channel resistance. In this experiment, a summing amplifier will be built and analyzed. The output voltage will be compared to the sum of the input voltages to ensure they match.

GOOD TO KNOW

Summing amplifiers are often used to combine multiple audio tracks. By varying the resistance value of the input resistors, the voltage gain for each input can be controlled.

Required Reading

Chapter 16 (Sec. 16-5) of *Electronic Principles,* 8th ed.

Equipment

1   signal generator
2   power supplies: ± 15 V
1   DMM (digital multimeter)
1   op amp: 741C
5   ½-W resistors: three 10 kΩ, 27 kΩ, 33 kΩ
2   switches: SPST
1   oscilloscope

Procedure

1. In the summing amplifier of Fig. 44-1, the source signal is 1 $V_{p\text{-}p}$ and 1 kHz. Calculate the voltage gain for each channel and record in Table 44-1. Then, calculate and record the peak-to-peak output voltage for the switch positions shown in the table.

2. Measure and record the resistor values. Build the circuit shown in Fig. 44-1. Use channel 1 of the oscilloscope to look at v_1 or v_2. Use channel 2 to look at v_{out}. Measure and record the output voltage for the switch positions shown in Table 44-2. Due to generator loading, the generator amplitude may need to be adjusted to ensure inputs of 1 $V_{p\text{-}p}$ on different switch positions.

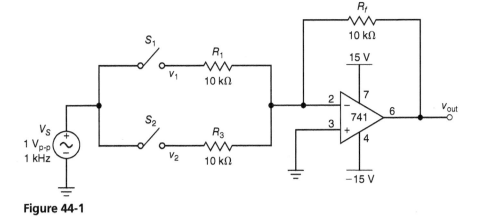

Figure 44-1

In other words, verify that v_1 and v_2 are the values listed in Table 44-2 before measuring v_{out}.

3. In Fig. 44-1, assume that R_1 is replaced with a 22 kΩ resistor. If the switches are closed and both inputs are 1 V_{p-p}, what is the output voltage? Record the answer here:

_____.

4. Replace R_1 with a 33 kΩ resistor. With both switches closed, adjust the generator amplitude so that $v_1 = v_2 = 1\ V_{p-p}$. Measure and record the output voltage here:

_____.

5. In Fig. 44-1, assume that $R_f = 27$ kΩ. If the switches are closed and both inputs are 1 V_{p-p}, what is the output voltage? Record the answer here:

_____.

6. Assume that $R_1 = R_2 = 10$ kΩ and $R_f = 27$ kΩ in Fig. 44-1. With the switches closed, calculate the output voltage for each input shown in Table 44-3.

7. Build the circuit with $R_1 = R_2 = 10$ kΩ and $R_f = 27$ kΩ. With both switches closed, adjust the generator amplitude so that v_1 equals the values listed in Table 44-3. Measure and record the output voltage.

ADDITIONAL WORK (OPTIONAL)

8. The circuit of Fig. 44-2 contains positive and negative clippers. With a sinusoidal source, two opposite-polarity half-wave signals drive the summing amplifier. Answer the following questions:

   a. If S_1 is closed and S_2 is open, what does the output of the summing amplifier look like?

   Answer = _____.

   b. With S_1 and S_2 closed, what does the upper 100 kΩ variable resistor do?

   Answer = _____.

   c. With S_1 and S_2 closed, what does the 10 kΩ variable resistor do?

   Answer = _____.

   d. What is the maximum voltage gain on channel 1?

   Answer = _____.

   e. What is the minimum voltage gain on channel 2?

   Answer = _____.

9. Build the circuit. Two jumper wires may be used in place of the SPST switches.

10. Adjust the signal generator to 20 V_{p-p} and 1 kHz. This will produce half-wave signals at the inputs to the switches. If the generator cannot produce 20 V_{p-p}, adjust the generator amplitude to its maximum.

11. Close S_1 and open S_2. Adjust the 10 kΩ variable resistance to maximum. With the oscilloscope, look at v_{out} while changing the upper 100 kΩ variable resistor over its entire range. Describe the output signal:

12. Adjust the upper 100 kΩ variable resistor to get an output voltage with a positive peak of 5 V.

13. Open S_1 and close S_2. With the oscilloscope, look at v_{out} while changing the lower 100 kΩ variable resistor over its entire range. Describe the output signal:

14. Adjust the lower 100 kΩ variable resistor to get an output voltage with a negative peak of −5 V.

15. Close S_1 and S_2. Describe the output signal:

16. Vary the 10 kΩ feedback resistor through its entire range. Describe the effect it has on the output signal:

17. Vary the upper and lower 100 kΩ potentiometers and observe the effect these have on each half-cycle of the output. Describe the output signal:

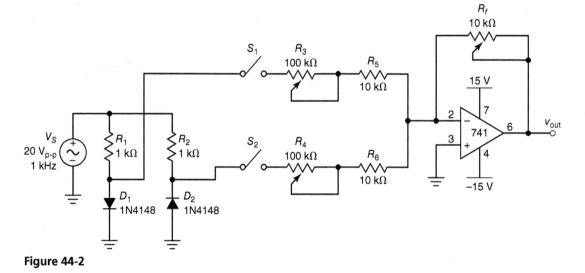

Figure 44-2

Experiment 44

Lab Partner(s) _____

PARTS USED Nominal Value	Measured Value
10 kΩ	
10 kΩ	
10 kΩ	
27 kΩ	
33 kΩ	

CALCULATIONS

TABLE 44-1. CALCULATIONS: A_{V1} = _____ AND A_{V2} = _____

S_1	S_2	v_1	v_2	v_{out}
Open	Open	0	0	
Open	Closed	0	1 V_{p-p}	
Closed	Open	1 V_{p-p}	0	
Closed	Closed	1 V_{p-p}	1 V_{p-p}	

TABLE 44-2. MEASUREMENTS: A_{V1} = _____ AND A_{V2} = _____

S_1	S_2	v_1	v_2	v_{out} Multisim	Actual
Open	Open	0	0		
Open	Closed	0	1 V_{p-p}		
Closed	Open	1 V_{p-p}	0		
Closed	Closed	1 V_{p-p}	1 V_{p-p}		

TABLE 44-3. S_1 AND S_2 CLOSED

v_1 or v_2	Calculated v_{out}	Measured v_{out} Multisim	Actual
0.5 V_{p-p}			
1 V_{p-p}			
1.5 V_{p-p}			
2 V_{p-p}			

Questions for Experiment 44

1. In Fig. 44-1, the voltage gain of each channel is: ( )
   (a) 0; (b) 1; (c) 2; (d) 10.
2. If each input voltage is 1 V_{p-p} in Fig. 44-1, the peak-to-peak output is: ( )
   (a) 0; (b) 1; (c) 2; (d) 10.

3. If R_1 is changed to 33 kΩ in Fig. 44-1, the peak-to-peak output is: ( )
    (a) 0;   (b) 1;   (c) 1.3;   (d) 5.4.
4. If R_f is changed to 27 kΩ in Fig. 44-1, the peak-to-peak output is: ( )
    (a) 0;   (b) 1;   (c) 1.3;   (d) 5.4.
5. If S_1 is open and S_2 is closed in Fig. 44-1, the peak-to-peak output is: ( )
    (a) 0;   (b) 1;   (c) 2;   (d) 10.
6. A summing amplifier has $v_1 = 1.5$ V dc, $v_2 = -2$ V, and unity voltage gain on   ( )
    each channel. Because of the phase inversion, the output voltage is:
    (a) 0.5 V;   (b) −0.5 V;   (c) 3.5 V;   (d) −3.5 V.
7. Many factors may contribute to the discrepancies between the calculated values of Table 44-1
    and the measured values of Table 44-2. But there is one factor that causes more error than
    any other. State what it is and why it produces the errors:

8. Suppose the following signals are present: the output of a microphone, the output of a CD
    player, and the audio output of a video camera. These three signals are to be mixed and fed
    into the microphone input of a laptop. Describe one way to do it:

9. Optional. Instructor's question.

10. Optional. Instructor's question.

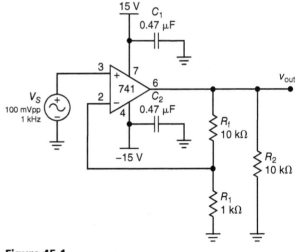

Experiment 45

VCVS Feedback

There are four basic types of negative feedback, depending on which input is used and which output quantity is sampled. VCVS feedback results in an almost perfect voltage amplifier, one with high input impedance, low output impedance, and stable voltage gain. The negative feedback also reduces nonlinear distortion and output offset voltage.

In this experiment, VCVS feedback will be investigated. First, the accuracy of the formula for closed-loop gain will be determined. Second, the stability of the voltage gain for different op amps will be determined. Third, the input offset voltages for different feedback resistors will be calculated and measured. Also included are troubleshooting and design.

GOOD TO KNOW

The characteristics of a VCVS negative feedback circuit make it a great choice when amplifying a small-input voltage.

Required Reading

Chapter 17 (Secs. 17-1 to 17-2) of *Electronic Principles*, 8th ed.

Equipment

1   signal generator
2   power supplies: ±15 V
9   ½-W resistors: two 1 kΩ, two 10 kΩ, 22 kΩ, 33 kΩ, 47 kΩ, 68 kΩ, 100 kΩ
3   op amps: 741C
2   capacitors: 0.47 μF
1   DMM (digital multimeter)
1   oscilloscope

Figure 45-1

Procedure

VOLTAGE AMPLIFIER

1. In Fig. 45-1, assume that R_f equals 10 kΩ. Calculate the closed-loop voltage gain. Record $A_{V(CL)}$ in Table 45-1.

2. Repeat Step 1 for the other values of R_f shown in Table 45-1.

3. Measure and record the resistor and capacitor values. Build the circuit shown in Fig. 45-1 with R_f equal to 10 kΩ. Set the signal generator to 1 kHz at 100 mV$_{p-p}$.

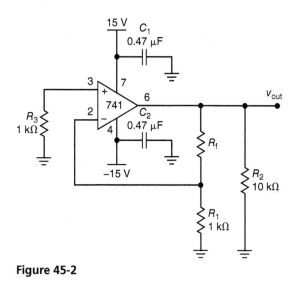

Figure 45-2

Measure v_{out}. Calculate the closed-loop voltage gain with $A_{V(CL)} = v_{out}/v_{in}$. Record this as the measured $A_{V(CL)}$.

4. Repeat Step 3 for the other values of R_f listed in Table 45-1.

STABLE VOLTAGE GAIN

5. Assume that R_f is 33 kΩ in Fig. 45-1. Calculate and record the closed-loop voltage gain (Table 45-2).
6. Build the circuit with R_f equal to 33 kΩ. Measure v_{out} and calculate $A_{V(CL)}$. Record this measured value in Table 45-2.
7. Repeat Steps 5 and 6 for the other 741Cs. If Multisim is being used, replace the LM741C op amp with LM725C and LM715C op amps for this step.

OUTPUT OFFSET VOLTAGE

8. Differences in the V_{BE} values and the input base currents imply there is a dc input offset voltage in Fig. 45-2. Assume that the total input offset voltage is 2 mV. Calculate and record the output offset voltage for each value of R_f listed in Table 45-3.

9. Build the circuit. Measure and record the output offset voltage for each value of R_f. (Even though the measured values may differ considerably from calculated values, the output offset voltage should increase with an increase in R_f.)

TROUBLESHOOTING

10. Assume that R_f is 100 kΩ in Fig. 45-1. For each trouble listed in Table 45-4, estimate the dc and peak-to-peak ac output voltage. Record the estimated values in Table 45-4.
11. Build the circuit with R_f equal to 100 kΩ. Insert each trouble into the circuit. Measure and record the dc and ac output voltages.

CRITICAL THINKING

12. Select a value of R_f in Fig. 45-1 to set up a closed-loop voltage gain of 40.
13. Build the circuit with the design value of R_f. Measure the closed-loop voltage gain. Record the design value for R_f and $A_{V(CL)}$ in Table 45-5.

ADDITIONAL WORK (OPTIONAL)

14. A 741C has a typical short-circuit output current of ±25 mA, or a peak-to-peak swing of 50 mA. Estimate the MPP for an 8 Ω speaker by multiplying 50 mA by 8 Ω. Record the answer here:

_____.

15. Calculate the speaker power using the MPP of the previous step.
16. A voltage follower using a 741C can drive a small 8 Ω speaker with enough power to produce an audible sound.
17. Build the circuit shown in Fig. 45-3.
18. Use channel 1 to measure the input voltage and use channel 2 to measure the output voltage. Use a sensitivity of 0.1 V/DIV on each channel.

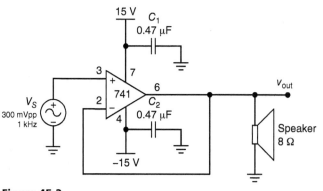

Figure 45-3

19. Adjust the peak-to-peak input voltage to 300 mV$_{p-p}$ and 1 kHz. A 1 kHz tone should be present. Vary the frequency, and the tone will change.
20. If the output waveform is clipped or distorted, reduce the input slightly until the clipping disappears.
21. Compare the input and output voltage waveforms. The two waveforms should be approximately the same since the circuit is a voltage follower. Since the load impedance is only 8 Ω, this means that a voltage follower is a stiff voltage source. The limitation on output power is the short-circuit current of a 741C, which is only ±25 mA. How could more power be delivered to the speaker?

22. Summarize what was learned in Steps 14 through 21.

Experiment 45

Lab Partner(s) _____

PARTS USED Nominal Value	Measured Value	PARTS USED Nominal Value	Measured Value	PARTS USED Nominal Value	Measured Value
1 kΩ		22 kΩ		100 kΩ	
1 kΩ		33 kΩ		0.47 μF	
10 kΩ		47 kΩ		0.47 μF	
10 kΩ		68 kΩ			

TABLE 45-1. CLOSED-LOOP VOLTAGE GAIN

R_f	Calculated $A_{V(CL)}$	Measured $A_{V(CL)}$	
		Multisim	Actual
10 kΩ			
22 kΩ			
47 kΩ			
68 kΩ			
100 kΩ			

TABLE 45-2. STABLE VOLTAGE GAIN

Op Amp	Calculated $A_{V(CL)}$	Measured $A_{V(CL)}$	
		Multisim	Actual
1			
2			
3			

TABLE 45-3. CLOSED-LOOP OUTPUT OFFSET VOLTAGE

R_f	Calculated $V_{oo(CL)}$	Measured $V_{oo(CL)}$	
		Multisim	Actual
10 kΩ			
22 kΩ			
47 kΩ			
68 kΩ			
100 kΩ			

TABLE 45-4. TROUBLESHOOTING

Trouble	Estimated V_{out}	Estimated v_{out}	Measured Multisim V_{out}	Measured Multisim v_{out}	Measured Actual V_{out}	Measured Actual v_{out}
R_f shorted						
R_f open						
R_1 shorted						
R_1 open						

TABLE 45-5. CRITICAL THINKING: $R_f = $ _____

Multisim $A_{V(CL)}$	Multisim % Error	Actual $A_{V(CL)}$	Actual % Error

Questions for Experiment 45

1. The calculated and measured $A_{V(CL)}$ of Table 45-1 were: ( )
   (a) extremely large; (b) very small; (c) close in value;
   (d) unpredictable.
2. The measured $A_{V(CL)}$ of Table 45-2 for all three 741Cs was: ( )
   (a) extremely large; (b) very small; (c) almost constant; (d) quite
   variable.
3. When R_f increases in Table 45-3, the closed-loop voltage gain increases and the ( )
   output offset voltage:
   (a) decreases; (b) increases; (c) stays the same; (d) none of the
   foregoing.
4. The closed-loop voltage gain of an amplifier with noninverting voltage feedback ( )
   is as stable as the:
   (a) supply voltage; (b) gain of the 741C; (c) load resistor;
   (d) feedback resistors.
5. If the input bias current is 80 nA in Fig. 45-2, the dc voltage across R_3 is: ( )
   (a) 80 μV; (b) 800 μV; (c) 2 mV; (d) −15 V.
6. What is the ac voltage at the inverting input of Fig. 45-1? Why?

TROUBLESHOOTING

7. When R_f is open or R_1 is shorted in Fig. 45-1, the output is clipped with a peak-to-peak value
   around 28 V. Explain why this happens.

8. When R_f is shorted or R_1 is open in Fig. 45-1, what does the closed-loop voltage gain equal? What is the name for this kind of circuit?

CRITICAL THINKING

9. If the voltage gain of an amplifier like the one shown in Fig. 45-1 needed to be within 2 percent, what would the design specifications be?

10. Optional. Instructor's question.

Negative Feedback

Always remember that there are four distinct types of negative feedback. Each type has different characteristics. VCVS feedback results in a voltage amplifier. VCIS feedback leads to a voltage-to-current converter. ICVS feedback results in a current-to-voltage converter. ICIS feedback leads to a current amplifier.

All four types of negative feedback reduce nonlinear distortion and output offset voltage. The noninverting types increase the input impedance, while the inverting types decrease it. The voltage feedback types decrease the output impedance, while the current feedback types increase the output impedance.

In this experiment, all four types of negative-feedback circuits using dc input and output voltages and currents will be built and analyzed.

GOOD TO KNOW

When measuring both voltage and current repeatedly, it is easy to have the DMM set up to measure current and then connect it in parallel with a component as if voltage was being measured. This action will cause the DMM's fuse to open.

Required Reading

Chapter 17 (Secs. 17-1 to 17-6) of *Electronic Principles,* 8th ed.

Equipment

1 signal generator
2 power supplies: ±15 V
6 ½-W resistors: two 1 kΩ, 2 kΩ, two 10 kΩ, 18 kΩ
1 potentiometer: 1 kΩ
1 op amp: 741C
2 capacitors: 0.47 μF
2 DMMs (digital multimeter): Two DMMs are preferred; however, the experiment can be done with a single DMM.

Procedure

VOLTAGE AMPLIFIER

1. For each dc input voltage listed in Table 46-1, calculate the dc ouput voltage of Fig. 46-1. Record the results.

2. Measure and record the resistor and capacitor values. Build the circuit shown in Fig. 46-1. Use one DMM on the input side and one on the output side. (If two DMMs are not available, the input voltage and output voltage can be measured separately.)

3. Adjust the potentiometer to get each dc input voltage listed in Table 46-1. Measure and record the output voltage.

VOLTAGE-TO-CURRENT CONVERTER

4. For each dc input voltage in Table 46-2, calculate the dc output current of Fig. 46-2. Record the results.

5. Build the circuit shown in Fig. 46-2. Use one DMM to measure the input voltage and one DMM to measure the output current. (If only one DMM is available, use a jumper in the place of the output ammeter when measuring the input voltage. When measuring output current, replace the jumper with an ammeter.)

6. Adjust the potentiometer to get an input voltage of 1 V. Read the output current and record the value in Table 46-2.

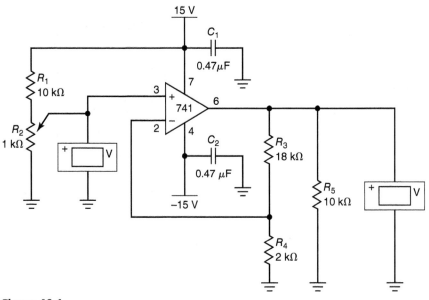

Figure 46-1

7. Repeat Step 6 for the remaining input voltages listed in Table 46-2.

CURRENT-TO-VOLTAGE CONVERTER

8. For each input current listed in Table 46-3, calculate the output voltage in Fig. 46-3. Record the results.
9. Build the circuit shown in Fig. 46-3.
10. Adjust the potentiometer to get an input current of 1 mA. Read the output voltage and record the value in Table 46-3.

11. Repeat Step 10 for the other input currents shown in Table 46-3.

CURRENT AMPLIFIER

12. For each input current listed in Table 46-4, calculate the output current in Fig. 46-4. Record the results.
13. Build the circuit shown in Fig. 46-4.
14. Adjust the potentiometer to get an input current of 0.1 mA. Record the output current in Table 46-4.

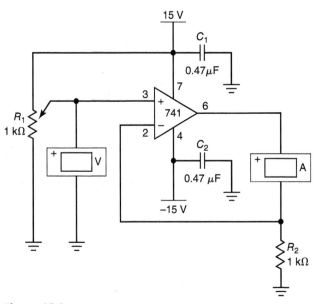

Figure 46-2

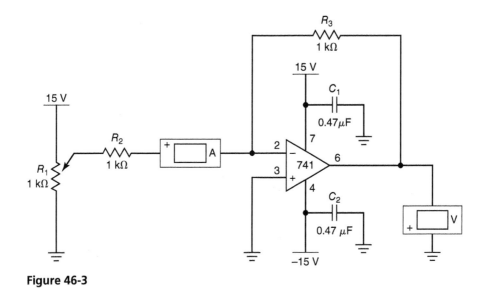

Figure 46-3

15. Repeat Step 14 for the remaining input currents of Table 46-4.

TROUBLESHOOTING

16. Ask the instructor to insert a trouble into any circuit.
17. Locate and repair the trouble. Record each trouble in Table 46-5.
18. Repeat Steps 16 and 17 as often as indicated by the instructor.

CRITICAL THINKING

19. The circuit of Fig. 46-2 has a transconductance of 100 μS. Redesign the circuit so that it has a g_m of 500 μS.
20. Build the redesigned circuit. Measure the output current for each voltage listed in Table 46-6.

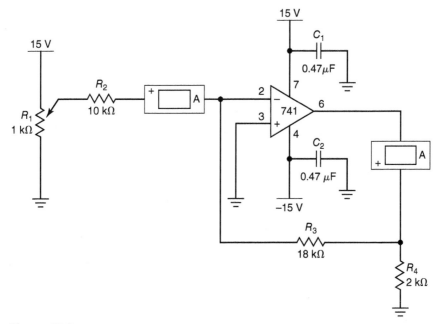

Figure 46-4

Experiment 46

Lab Partner(s) _____

PARTS USED Nominal Value	Measured Value	PARTS USED Nominal Value	Measured Value	PARTS USED Nominal Value	Measured Value
1 kΩ		10 kΩ		1 kΩ Pot	
1 kΩ		10 kΩ		0.47 μF	
2 kΩ		18 kΩ		0.47 μF	

TABLE 46-1. VCVS FEEDBACK

V_{in}	Calculated V_{out}	Measured V_{out}	
		Multisim	Actual
0.1 V			
0.2 V			
0.3 V			
0.4 V			
0.6 V			
0.8 V			
1 V			

TABLE 46-2. VCIS FEEDBACK

V_{in}	Calculated I_{out}	Measured I_{out}	
		Multisim	Actual
1 V			
2 V			
3 V			
4 V			
6 V			
8 V			
10 V			

TABLE 46-3. ICVS FEEDBACK

I_{in}	Calculated V_{out}	Measured V_{out}	
		Multisim	Actual
1 mA			
2 mA			
3 mA			
4 mA			
6 mA			
8 mA			
10 mA			

TABLE 46-4. ICIS FEEDBACK

I_{in}	Calculated I_{out}	Measured I_{out}	
		Multisim	Actual
0.1 mA			
0.2 mA			
0.3 mA			
0.4 mA			
0.6 mA			
0.8 mA			
1 mA			

TABLE 46-5. TROUBLESHOOTING

Trouble	Description
1	
2	
3	

TABLE 46-6. CRITICAL THINKING

V_{in}	Multisim		Actual	
	I_{out}	% Error	I_{out}	% Error
1 V				
2 V				
3 V				
4 V				
6 V				
8 V				
10 V				

Questions for Experiment 46

1. The voltage gain of Table 46-1 is closest to: ( )
   (a) 1;   (b) 5;   (c) 10;   (d) 20.
2. The transconductance of Table 46-2 is approximately: ( )
   (a) 100 μS;   (b) 300 μS;   (c) 750 μS;   (d) 1000 μS.
3. The transresistance of Table 46-3 is approximately: ( )
   (a) 100 Ω;   (b) 1 kΩ;   (c) 10 kΩ;   (d) 100 kΩ.
4. The current gain of Table 46-4 is closest to: ( )
   (a) 1;   (b) 10;   (c) 100;   (d) 1000.
5. The stability or accuracy of any of the feedback circuits in this experiment depends ( )
   primarily on the:
   (a) supply voltage;   (b) 741C;   (c) DMM;   (d) tolerance of feedback resistors.
6. What was learned from this experiment? List at least two ideas that seemed to stand out.

TROUBLESHOOTING

7. A 10 kΩ resistor was mistakenly used for the feedback resistor of Fig. 46-2. How will this affect the circuit performance?

8. The negative supply voltage of Fig. 46-3 is not connected to the op amp. What are the symptoms of this trouble?

CRITICAL THINKING

9. An electronic VOM is being designed. Which of the basic feedback circuits would be used to measure voltage? Which would be used for measuring current?

10. Optional. Instructor's question.

Gain-Bandwidth Product

When working with an op amp, remember that the gain-bandwidth product is a constant. This means the product of closed-loop voltage gain and bandwidth equals the unity-gain frequency of the op amp. Stated another way, it means that voltage gain can be traded for bandwidth. For instance, if the voltage gain is reduced by a factor of 2, the bandwidth will double.

In this experiment, the bandwidth for different voltage gains will be determined through calculation and actual measurement. This will confirm that the gain-bandwidth product is a constant.

GOOD TO KNOW

If the voltage gain required is greater than the gain-bandwidth product allows, multiple stages can be used to increase the voltage gain. Remember, the overall A_V is the product of the A_V for each stage.

Required Reading

Chapter 17 (Sec. 17-7) of *Electronic Principles,* 8th ed.

Equipment

1   function generator
2   power supplies: ±15 V
7   ½-W resistors: 100 Ω, 4.7 kΩ, 6.8 kΩ, 10 kΩ, 22 kΩ, 33 kΩ, 47 kΩ
1   op amp: 741C
2   capacitors: 0.47 μF
1   DMM (digital multimeter)
1   oscilloscope
1   frequency counter

Procedure

CALCULATING VOLTAGE GAIN AND BANDWIDTH

1. For each value of R_1 listed in Table 47-1, calculate the closed-loop voltage gain of Fig. 47-1. Record all answers.

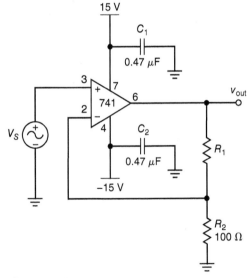

Figure 47-1

2. The typical gain-bandwidth product (same as f_{unity}) of a 741C is 1 MHz. Calculate and record the closed-loop cutoff frequency for each R_1 listed in Table 47-1.

3. Measure and record the resistor and capacitor values. Build the circuit shown in Fig. 47-1 with R_1 equal to 4.7 kΩ. Look at the output signal with an oscilloscope. With the input frequency at 100 Hz, adjust the signal level to get an output of 5 V_{p-p}.

4. Measure the peak-to-peak input voltage. Calculate and record $A_{V(CL)}$ as a measured quantity (Table 47-1).

5. Measure and record the upper cutoff frequency.

6. Repeat Steps 3 through 5 for the other values of R_1 in Table 47-1.

MEASURING RISETIME TO GET BANDWIDTH

7. Build the circuit shown in Fig. 47-1 with an R_1 of 4.7 kΩ and a square-wave output from the function generator.

8. Set the frequency to approximately 5 kHz, and adjust the signal level to get an output voltage of 5 V_{p-p}.

9. Measure the risetime and record in Table 47-2. Calculate and record $f_{2(CL)}$.

10. Repeat Steps 7 through 9 for the other values of R_1. (Note: An input frequency less than 5 kHz will be needed as the value of R_1 increases. Reduce the frequency as needed to get an accurate risetime measurement.)

TROUBLESHOOTING

11. Estimate the risetime in Fig. 47-1 for each trouble listed in Table 47-3. Record the results.

12. Insert each trouble. Measure and record the risetime.

CRITICAL THINKING

13. Select a value of R_1 in Fig. 47-1 to get a bandwidth of 35 kHz. (Use a 741C.)

14. Build the circuit shown in Fig. 47-1 with a selected value of R_1. Measure the voltage gain and risetime. Calculate the bandwidth. Record all quantities listed in Table 47-4.

Experiment 47

Lab Partner(s) _____

PARTS USED Nominal Value	Measured Value	PARTS USED Nominal Value	Measured Value	PARTS USED Nominal Value	Measured Value
100 Ω		10 kΩ		47 kΩ	
4.7 kΩ		22 kΩ		0.47 μF	
6.8 kΩ		33 kΩ		0.47 μF	

TABLE 47-1. GAIN AND CRITICAL FREQUENCY

| | Calculated | | Measured | | | |
| | | | Multisim | | Actual | |
R_1	$A_{V(CL)}$	$f_{2(CL)}$	$A_{V(CL)}$	$f_{2(CL)}$	$A_{V(CL)}$	$f_{2(CL)}$
4.7 kΩ						
6.8 kΩ						
10 kΩ						
22 kΩ						
33 kΩ						
47 kΩ						

TABLE 47-2. RISETIME

| | Multisim | | Actual | |
R_1	Measured T_R	Experimental $f_{2(CL)}$	Measured T_R	Experimental $f_{2(CL)}$
4.7 kΩ				
6.8 kΩ				
10 kΩ				
22 kΩ				
33 kΩ				
47 kΩ				

TABLE 47-3. TROUBLESHOOTING

| | | Measured T_R | |
Trouble	Estimated T_R	Multisim	Actual
R_1 shorted			
No +15 V supply			
100 Ω open			

TABLE 47-4. CRITICAL THINKING: $R_1 =$ _____

Multisim				Actual			
$A_{V(CL)}$	T_R	$f_{2(CL)}$	% Error	$A_{V(CL)}$	T_R	$f_{2(CL)}$	% Error

Questions for Experiment 47

1. The measured data of Table 47-1 indicate that the product of gain and bandwidth is: ( )
   (a) 1 MHz; **(b)** approximately constant; **(c)** variable; **(d)** none of the foregoing.
2. The largest value of R_1 in Table 47-2 produces the: ( )
   (a) smallest T_R; **(b)** largest T_R; **(c)** smallest voltage gain; **(d)** none of the foregoing.
3. In Fig. 47-1, an increase in voltage gain leads to a(n): ( )
   (a) decrease in bandwidth; **(b)** increase in bandwidth; **(c)** loss of supply voltage; **(d)** smaller output voltage.
4. If an op amp has a higher f_{unity}, the bandwidth will be higher for a given: ( )
   (a) supply voltage; **(b)** voltage gain; **(c)** output voltage; **(d)** MPP value.
5. To increase the bandwidth of a circuit like Fig. 47-1: ( )
   (a) decrease the voltage gain; **(b)** increase the supply voltage; **(c)** decrease the f_{unity}; **(d)** increase the output voltage.
6. Why is it important to know that the gain-bandwidth product is constant?

TROUBLESHOOTING

7. Suppose one of the bypass capacitors of Fig. 47-1 shorts out. What symptoms would be present?

8. There is no dc or ac output voltage in a circuit like Fig. 47-1. Name three possible causes.

CRITICAL THINKING

9. When designing an amplifier to have as fast a risetime as possible, should the op amp have a low or a high f_{unity}? Why?

10. Optional. Instructor's question.

Linear IC Amplifiers

Linear op-amp circuits preserve the shape of the input signal. If the input is sinusoidal, the output will be sinusoidal. Two basic voltage amplifiers are possible: the noninverting amplifier and the inverting amplifier. The inverting amplifier consists of a source resistance cascaded with a current-to-voltage converter. As discussed in the textbook, the closed-loop voltage gain equals the ratio of the feedback resistance to the source resistance.

In this experiment, both types of voltage amplifiers will be built and tested. A noninverter/inverter with a single adjustment that allows the voltage gain to be varied will also be built and tested.

GOOD TO KNOW

The output voltage will decrease to 0.707 of v_{midband} at the upper corner frequency ($f_{2(CL)}$). Example: If $v_{\text{midband}} = 5V_{\text{p-p}}$, at $f_{2(CL)}$ $v_{\text{out}} = 5\ V_{\text{p-p}} \times 0.707$ or $3.535\ V_{\text{p-p}}$.

Required Reading

Chapter 18 (Secs. 18-1 to 18-6) of *Electronic Principles*, 8th ed.

Equipment

- 1 function generator
- 2 power supplies: ± 15 V
- 13 ½-W resistors: 100 Ω, two 1 kΩ, 1.1 kΩ, two 6.8 kΩ, 10 kΩ, 47 kΩ, 68 kΩ, 100 kΩ, 220 kΩ, 330 kΩ, 470 kΩ
- 1 potentiometer: 1 kΩ
- 1 op amp: 741C
- 4 capacitors: two 0.47 μF, two 1 μF
- 1 DMM (digital multimeter)
- 1 oscilloscope
- 1 frequency counter

Procedure

SINGLE-SUPPLY NONINVERTING AMPLIFIER

1. Assume a typical f_{unity} of 1 MHz for the 741C of Fig. 48-1. Calculate $A_{V(CL)}$ and $f_{2(CL)}$. Also calculate the input, output, and bypass critical frequencies. Estimate the MPP value. Record all answers in Table 48-1.

2. Measure and record the resistor and capacitor values. Build the circuit shown in Fig. 48-1. Adjust the function generator to 100 mV$_{\text{p-p}}$ at 1 kHz. Measure and record $A_{V(CL)}$.

3. Measure and record the upper cutoff frequency. (Try both the sine- and the square-wave methods.)

4. Measure and record the lower cutoff frequency.

5. Measure and record the MPP value.

INVERTING AMPLIFIER

6. For each R_3 value of Table 48-2, calculate $A_{V(CL)}$ and $f_{2(CL)}$ in Fig. 48-2.

7. Connect the circuit with R_3 equal to 47 kΩ. Set the input frequency to 100 Hz. Adjust the signal level to get an output of 5 V$_{\text{p-p}}$.

8. Measure v_{in}. Calculate and record $A_{V(CL)}$ as a measured quantity.

9. Measure and record $f_{2(CL)}$.

10. Repeat Steps 7 through 9 for other R_3 values in Table 48-2.

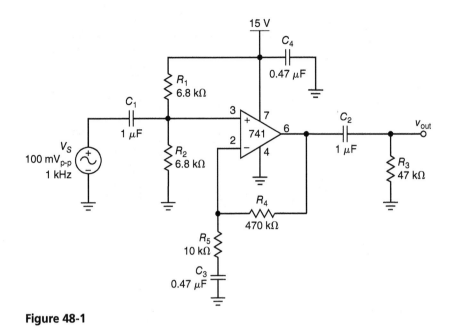

Figure 48-1

NONINVERTER/INVERTER

11. Calculate the maximum noninverting and inverting voltage gains for the circuit of Fig. 48-3. Record in Table 48-3.
12. Build the circuit shown in Fig. 48-3.
13. Look at the output signal with an oscilloscope. Vary the potentiometer and observe what happens.
14. Measure the maximum noninverting and inverting voltage gains. Record the data in Table 48-3.

TROUBLESHOOTING

15. For each trouble listed in Table 48-4, estimate and record the dc voltage at pin 6 (Fig. 48-1).
16. Insert each trouble into the circuit. Measure and record the dc voltage at pin 6.

CRITICAL THINKING

17. Select new values for C_1 and C_3 to get a lower cutoff frequency in Fig. 48-1 that is less than 20 Hz.
18. Build the circuit. Measure and record the lower cutoff frequency. Record all quantities listed in Table 48-5.

APPLICATION (OPTIONAL) — MICROPHONE PREAMP

19. In Fig. 48-4, the MCM 35-4955 is a microphone. It has a variable resistance that changes when sound waves strike it. The changing resistance produces an ac voltage, which is an electrical representation of the sound.
20. Build the circuit shown in Fig. 48-4.
21. The amplitude of the output signal will depend on the loudness and proximity of the sound. Observe

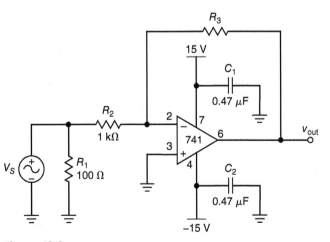

Figure 48-2

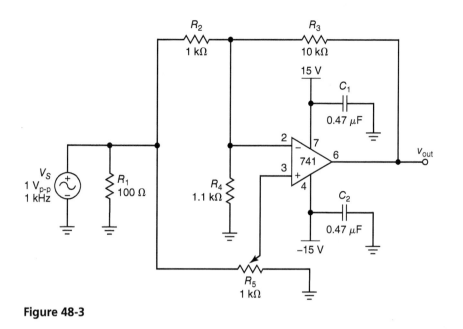

Figure 48-3

the output signal of the op amp with an oscilloscope. Set the sensitivity to 20 mV/DIV and the time base to 10 ms/DIV.

22. Watch the output signal while someone speaks into the microphone. A signal will vary as the sound varies. Experiment with different sensitivities from 10 to 200 mV/DIV as someone speaks into the microphone. Also try different time bases from 1 to 20 ms/DIV. (What should one say when speaking into the microphone? To quote the immortal words of Alexander Graham Bell's first phone call: "Come here, Watson, I need you.")

ADDITIONAL WORK (OPTIONAL) — REVERSIBLE AND ADJUSTABLE GAIN

23. The circuit of Fig. 48-5 allows the voltage gain to be varied continuously from −1 to 0, and then from 0 to +1.
24. Build the circuit shown in Fig. 48-5.
25. Use the oscilloscope to look at the input signal (channel 1) and the output signal (channel 2). Set both sensitivities to 0.5 V/DIV. Adjust the generator amplitude to get an output signal of 1 V_{p-p}.
26. Observe the input and signals while the 10 kΩ potentiometer is varied through its entire range. Describe what the circuit does:

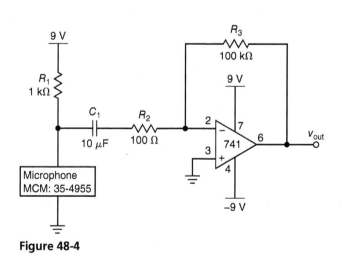

Figure 48-4

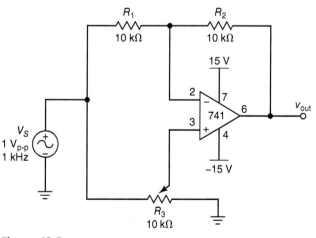

Figure 48-5

265

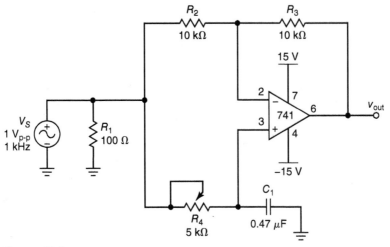

Figure 48-6

PHASE SHIFTER

27. Build the circuit shown in Fig. 48-6.
28. Use channel 1 of the oscilloscope for the input signal and use channel 2 for the output signal. Set both sensitivities to 0.5 V/DIV. Adjust the generator amplitude to get an output voltage of 1 V_{p-p} and a frequency of 1 kHz.
29. Observe the input and output signals while varying the resistor through its range. Describe what the circuit does:

WHEATSTONE BRIDGE AND DIFFERENTIAL AMPLIFIER

30. Build the circuit shown in Fig. 48-7.
31. Use a DMM to measure and record the differential voltage out of the bridge:

    $v_{in} =$ _____.

32. Use a DMM to measure the output voltage of the op amp:

    $v_{out} =$ _____.

33. Calculate the differential voltage gain:

    $A_V =$ _____.

34. Place a short between points *A* and *B*. In this case, the input is a common-mode signal because $v_1 = v_2$. Use the DMM to measure v_1 with respect to ground:

    $v_{in(CM)} =$ _____.

35. Use a DMM to measure the output voltage of the op amp:

    $v_{out(CM)} =$ _____.

36. Calculate the common-mode voltage gain and record its value:

    $A_{V(CM)} =$ _____.

37. Calculate the common-mode rejection ratio:

    CMRR = _____.

38. What was learned in Steps 31 through 37?

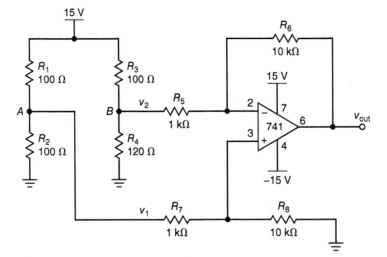

Figure 48-7

Experiment 48

Lab Partner(s) _____

PARTS USED Nominal Value	Measured Value	PARTS USED Nominal Value	Measured Value	PARTS USED Nominal Value	Measured Value
100 Ω		10 kΩ		470 kΩ	
1 kΩ		47 kΩ		1 kΩ Pot	
1 kΩ		68 kΩ		0.47 μF	
1.1 kΩ		100 kΩ		0.47 μF	
6.8 kΩ		220 kΩ		1 μF	
6.8 kΩ		330 kΩ		1 μF	

TABLE 48-1. NONINVERTING AMPLIFIER

	Calculated	Measured	
		Multisim	Actual
$A_{V(CL)}$			
$f_{2(CL)}$			
f_{in}			
f_{out}			
f_{BY}			
MPP			

TABLE 48-2. INVERTING AMPLIFIER

	Calculated		Measured			
			Multisim		Actual	
R_3	$A_{V(CL)}$	$f_{2(CL)}$	$A_{V(CL)}$	$f_{2(CL)}$	$A_{V(CL)}$	$f_{2(CL)}$
47 kΩ						
68 kΩ						
100 kΩ						
220 kΩ						
330 kΩ						
470 kΩ						

TABLE 48-3. NONINVERTER/INVERTER

Calculated		Measured			
		Multisim		Actual	
$A_{V(non)}$	$A_{V(inv)}$	$A_{V(non)}$	$A_{V(inv)}$	$A_{V(non)}$	$A_{V(inv)}$

TABLE 48-4. TROUBLESHOOTING

Trouble	Estimated Pin 6 Voltage	Measured Pin 6 Voltage	
		Multisim	Actual
R_1 open			
R_1 shorted			
R_2 open			
R_2 shorted			
C_1 open			

TABLE 48-5. CRITICAL THINKING: $C_1 =$ _____ ; $C_3 =$ _____

	Multisim	% Error	Actual	% Error
$f_{1(CL)}$				

Questions for Experiment 48

1. The MPP value in Table 48-1 is closest to: ( )
   (a) 1 V;   (b) 7.5 V;   (c) 12.5 V;   (d) 20 V.
2. The product of voltage gain and bandwidth in Table 48-2 is: ( )
   (a) approximately constant;   (b) small;   (c) 100;   (d) 20 kHz.
3. The noninverter/inverter of Fig. 48-3 has a noninverting voltage gain of   ( )
   approximately:
   (a) 1;   (b) 10;   (c) 100;   (d) 1000.
4. The bypass capacitor of Fig. 48-1 sets up a cutoff frequency of approximately: ( )
   (a) 3.39 Hz;   (b) 33.9 Hz;   (c) 46.8 Hz;   (d) 63 Hz.
5. With the inverter of Fig. 48-2, the bandwidth can be increased by: ( )
   (a) decreasing the supply voltage;   (b) decreasing the voltage gain;
   (c) increasing the value of R;   (d) eliminating the bypass capacitors.
6. Explain how the inverter of Fig. 48-2 works.

TROUBLESHOOTING

7. In Fig. 48-1, what happens to the dc voltage at pin 6 when R_1 is shorted? Why?

8. Suppose the bypass capacitor C_3 opens in Fig. 48-1. What kind of dc and ac symptoms will be present at the output of the op amp?

CRITICAL THINKING

9. Explain why the values for C_1 and C_3 were selected for the design.

10. Optional. Instructor's question.

Single Power Supply Audio Op-Amp Application

In this experiment, an audio amplifier utilizing an operational amplifier in a single power supply configuration with diode input protection will be constructed from discrete components and powered by a 12- to 15-volt dc variable power supply. The op-amp circuit uses a diode protection circuit (D_1 and D_2) to prevent the audio amplifier from being damaged from too large of an input signal. The 9-volt regulator provides the $+9$ V to pin 7 of the op amp. Pin 4 of the op amp is connected to ground instead of a negative power supply. R_3 and R_4 form a voltage divider, providing $+4.5$ V to the noninverting input of the op amp. The amplified output signal will be centered on the $+4.5$ Vdc, as seen at test point D. C_4 will allow the signal to pass through to the load resistor while blocking the 4.5 Vdc level, as seen at test point E.

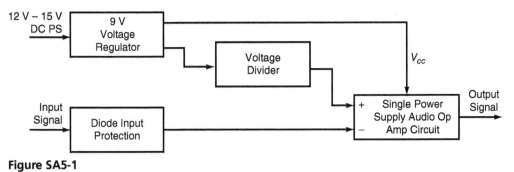

Figure SA5-1

In this experiment, the use of an oscilloscope to trace a signal through a single power supply op-amp circuit will be introduced. This experiment will also provide an opportunity to troubleshoot a single power supply op-amp circuit.

GOOD TO KNOW

Since most components are soldered to a printed circuit board, voltage is the easiest circuit parameter to measure.

Required Reading

Chapters 16 to 18 of *Electronic Principles,* 8th ed.

Equipment

1 variable dc power supply (+12 V to +15 V)
1 voltage regulator: LM7809 (or equivalent)
1 op amp: LM741
2 diodes: 1N914 or 1N4148
6 ½-W resistors: 100 Ω, 220 Ω, three 10 kΩ, 293 kΩ
5 capacitors: 0.1 μF, 0.33 μF, 1 μF, 33 μF, 47 μF (25-V rating or higher)
1 5 kΩ potentiometer
1 DMM (digital multimeter)
1 oscilloscope

Procedure

POWER SUPPLY CIRCUIT ANALYSIS

1. Sketch the pinout of the LM7809 9V voltage regulator. Using a DMM, measure and record the value of the resistors and capacitors. Change the DMM mode to Diode Test, and measure and record the voltage across both silicon diodes when forward and reverse biased.

2. Build the op-amp circuit as shown in Fig. SA5-2. (A 290 kΩ resistor in series with a 5 kΩ potentiometer may be used to obtain the 293 kΩ resistance. Be sure to tie the center wiper to one of the outside legs of the potentiometer.)

3. Use a variable dc power supply set to 13 volts.

4. Based on the material in the textbook and the data sheet for the LM7809, predict the voltages that would be present at test points A and B. Record the predictions in Table SA5-1.

5. R_3 and R_4 form a voltage divider. Based on the predicted dc voltage value at test point B, calculate the expected dc voltage at test point D. Record the calculated values in Table SA5-1.

6. Using a DMM, measure the dc voltages present at test points A, B and D. Record the measured values in Table SA5-1.

VOLTAGE GAIN ANALYSIS

7. Based on the values shown for R_1 and R_2, calculate the voltage gain for the circuit. Record the calculated voltage gain in Table SA5-2.

8. Using the oscilloscope, observe and measure the dc voltage and ac peak-to-peak signal voltage at test point C (V_S) with channel 1. Using the oscilloscope, observe and measure the dc voltage and ac peak-to-peak signal voltage present at test point F (V_{RL}) with channel 2. Record the measured values in Table SA5-2. (*Note:* Use dc coupling for the dc voltage measurement and ac coupling for the ac voltage measurement.)

9. Using the measured values for V_S (TP C) and V_{RL} (TP F), calculate the actual voltage gain. Record the actual voltage gain in Table SA5-2.

TROUBLESHOOTING

10. Assume the circuit has the faults listed in Table SA5-3. Record the anticipated voltages at the test points listed for each assumed fault.

11. Insert the faults listed in Table SA5-3 into the circuit and verify the predicted results by measuring the voltages present at the test points listed.

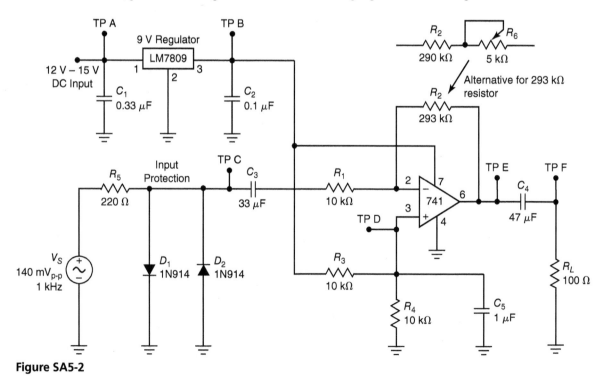

Figure SA5-2

CRITICAL THINKING

12. R_3 and R_4 form a voltage divider. They provide approximately $V_{CC}/2$ (+4.5 V) as the input voltage for pin 3, the noninverting input. The amplified signal is centered on the +4.5 V at test point E. Using the oscilloscope, while observing the input signal at test point C with channel 1, measure the ac peak-to-peak signal voltage at test point E with channel 2 and test point F with channel 3. (If only two channels are available, view the signals present at test points C and E.) Slowly increase the amplitude of the function generator until the signal at test point E begins to clip. Explain why the signal clips. What is limiting the amplitude of the output signal?

13. If a 1 kΩ resistor was used in place of the 10 kΩ for R_4, how would that affect the circuit?

14. Could the voltage regulator be replaced with a +9 V battery? Draw the schematic with the battery in place of the voltage regulator. Optional: Replace the voltage regulator with the +9 V battery and verify the circuit's operation.

System Application 5

Lab Partner(s) _____

PARTS USED Nominal Value	Measured Value	LM7809 PINOUT		
100 Ω				
220 Ω				
10 kΩ				
10 kΩ				
10 kΩ				
293 kΩ				
0.1 μF				
0.33 μF				
1 μF			V_F	V_R
33 μF		1N914		
47 μF		1N914		

TABLE SA5-1 DC VOLTAGE MEASUREMENTS

Test Point	Calculated / Predicted	Multisim	Actual
A			
B			
D			

TABLE SA5-2 VOLTAGE GAIN MEASUREMENTS

Calculated	Multisim		Actual	
A_V	Voltages	A_V	Voltages	A_V
	TP C		TP C	
	TP F		TP F	

TABLE SA5-3 TROUBLESHOOTING

Trouble	TP B	TP C	TP D	TP F
C_5 is shorted				
R_2 shorted				
LM7809 fails				
D_1 shorts				
R_1 opens				

Questions for System Application 5

1. R_3 and R_4 form a voltage divider to provide _____ at test point D.
   (a) +4.5 V;   (b) +3.2 V;   (c) +5.2 V;   (d) +7.8 V.
2. The maximum amplitude of the output signal is limited to _____.
   (a) $\approx V_{CC}/2$;   (b) $2 \times V_{CC}$;   (c) V_{CC};   (d) $V_{CC} \times 0.707$.
3. Diodes D_1 and D_2 limit the input signal to approximately _____.
   (a) +1.4 V;   (b) +0.7 V;   (c) +100 mV;   (d) +11.5 V.
4. The calculated A_V for the circuit is _____.
   (a) 38.6;   (b) 10.8;   (c) 29.3;   (d) 11.7.
5. According to the data sheet for the LM7809, a minimum of _____ is required to obtain +9 V on the output.
   (a) +15 V;   (b) +11.5 V;   (c) +10 V;   (d) +13.8 V.
6. The purpose of capacitor C_4 is to _____.
   (a) smooth out the ac ripple;   (b) eliminate any hum;   (c) block the dc voltage while allowing the ac voltage to pass through;   (d) none of the answers listed.
7. If C_5 was shorted, the voltage present at TP D would be _____.
   (a) 5 V;   (b) 4.5 V;   (c) 9 V;   (d) 0 V.
8. If D_2 was shorted, the signal present at TP C would be _____.
   (a) 0 V;   (b) 140 mV$_{p-p}$;   (c) 0.7 V$_P$;   (d) 70 mV$_{p-p}$.
9. If C_3 was open, the signal present at TP F would be _____.
   (a) 4 V$_{p-p}$;   (b) 140 mV$_{p-p}$;   (c) 0 V;   (d) 4.5 V$_{p-p}$.
10. The purpose of capacitors C_1 and C_2 is to _____.
    (a) remove ac ripple;   (b) avoid voltage spikes;   (c) prevent oscillations;
    (d) limit current fluctuation.

Current Boosters and Controlled Current Sources

For a given input voltage, the load current of a VCVS amplifier depends on the load resistance. The smaller the load resistance, the larger the current. A typical op amp like the 741 cannot drive very small load resistances because its maximum output current is 25 mA. When a load requires more current than this, a current booster can be added to the output of the op amp. In this experiment, a voltage follower with a current booster will be built and its operation analyzed.

The voltage-controlled current source uses VCIS negative feedback. Because of this, a given input voltage sets up a constant load current. With a VCIS amplifier, the load resistance can be changed without changing the load current. In this experiment, a voltage-controlled current source will be built and its operation analyzed.

GOOD TO KNOW

Power transistors are often used to deliver current to the load when the current requirements exceed the original circuit.

Required Reading

Chapter 18 (Secs. 18-7 and 18-8) of *Electronic Principles,* 8th ed.

Equipment

1 signal generator
2 power supplies: ± 15 V
10 ½-W resistors: 10 Ω, 33 Ω, 100 Ω, 330 Ω, three
   1 kΩ, 1.8 kΩ, 2.2 kΩ, 10 kΩ
1 potentiometer: 1 kΩ
3 transistors: 2N3904, 2N3906, 2N3055
2 op amps: 741C

4 capacitors: 0.47 μF
1 DMM (digital multimeter)
1 oscilloscope

Procedure

VOLTAGE FOLLOWER WITHOUT CURRENT BOOSTER

1. Measure and record the resistor and capacitor values. Build the circuit shown in Fig. 49-1 with $R_L = 1$ kΩ.
2. Measure the input voltage to pin 3 and record the value in Table 49-1.
3. Measure and record the load voltage and load current for each of the load resistances shown in Table 49-1.

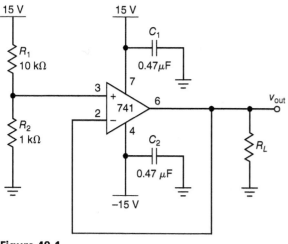

Figure 49-1

4. Record any observations about the load current for smaller resistances:

VOLTAGE FOLLOWER WITH CURRENT BOOSTER

5. Add a current booster to the voltage follower, as shown in Fig. 49-2.
6. Repeat Steps 2 and 3 for Table 49-2.
7. What was learned about an op-amp circuit when used without a current booster? With a current booster?

VOLTAGE-CONTROLLED CURRENT SOURCE

8. In Fig. 49-3, assume that R_L is 100 Ω and calculate the load current for each input voltage listed in Table 49-3.

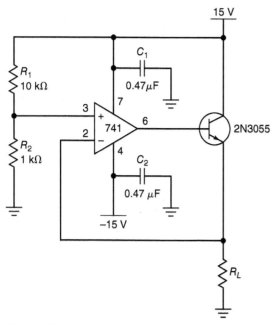

Figure 49-2

278

9. Assume that R_L is 2.2 kΩ and calculate the load current for each input voltage in Table 49-4.
10. Build the circuit shown in Fig. 49-3 with R_L equal to 100 Ω.
11. Adjust the input voltage to each value in Table 49-3. Measure and record the load current.
12. Change R_L to 2.2 kΩ. Measure the load current for each input voltage of Table 49-4.

TROUBLESHOOTING

13. In Fig. 49-3, assume an R_L of 100 Ω and a v_{in} of +5 V. Estimate and record the load current for each trouble of Table 49-5.
14. Insert each trouble into the circuit. Measure and record the load current.

CRITICAL THINKING

15. Redesign the circuit of Fig. 49-3 to get a load current of approximately 2.25 mA when v_{in} is +5 V.
16. Build the circuit with the selected values. Measure the load current with a v_{in} of +5 V. Record quantities listed in Table 49-6.

ADDITIONAL WORK (OPTIONAL)

17. Build the Howland current source shown in Fig. 49-4.
18. Calculate the regulated load current for an input voltage of 3 V. Record the answer:
    $I_L =$ _____.
19. Use a DMM for voltage and current measurements. Adjust the potentiometer to get an input voltage of 3 V. Measure the load current for each of the following load resistances:

    10 Ω: $I_L =$ _____.

    100 Ω: $I_L =$ _____.

    1 kΩ: $I_L =$ _____.
20. Increase the input voltage to 9 V.
21. Repeat the measurements:

    10 Ω: $I_L =$ _____.

    100 Ω: $I_L =$ _____.

    1 kΩ: $I_L =$ _____.
22. Use −15 V for the potentiometer instead of +15 V. Then measure the load current for the following load resistances:

    10 Ω: $I_L =$ _____.

    100 Ω: $I_L =$ _____.

    1 kΩ: $I_L =$ _____.
23. Describe the action of a Howland current source. Why is it more useful than the current source in Fig. 49-3?

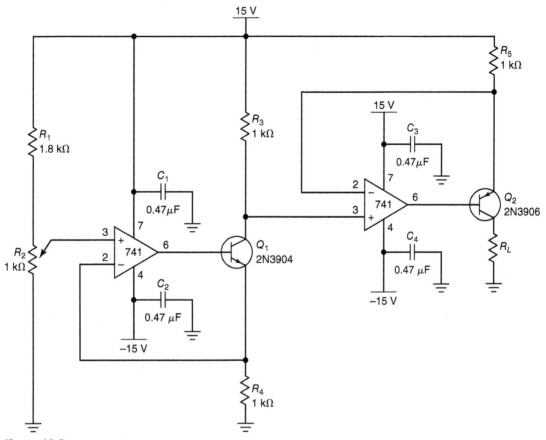

Figure 49-3

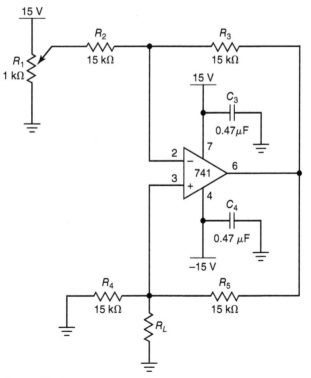

Figure 49-4

Experiment 49

Lab Partner(s) _____

PARTS USED Nominal Value	Measured Value	PARTS USED Nominal Value	Measured Value	PARTS USED Nominal Value	Measured Value
10 Ω		1 kΩ		1 kΩ Pot	
33 Ω		1 kΩ		0.47 μF	
100 Ω		1.8 kΩ		0.47 μF	
330 Ω		2.2 kΩ		0.47 μF	
1 kΩ		10 kΩ		0.47 μF	

TABLE 49-1. VOLTAGE FOLLOWER (WITHOUT CURRENT BOOSTER): V_{in} = _____

R_L	Multisim		Actual	
	V_L	I_L	V_L	I_L
1 kΩ				
330 Ω				
100 Ω				
33 Ω				
10 Ω				

TABLE 49-2. VOLTAGE FOLLOWER (WITH CURRENT BOOSTER): V_{in} = _____

R_L	Multisim		Actual	
	V_L	I_L	V_L	I_L
1 kΩ				
330 Ω				
100 Ω				
33 Ω				
10 Ω				

TABLE 49-3. VOLTAGE-CONTROLLED CURRENT SOURCE: R_L = 100 Ω

V_{in}	Calculated I_L	Measured I_L	
		Multisim	Actual
0 V			
1 V			
2 V			
3 V			
4 V			
5 V			

TABLE 49-4. VOLTAGE-CONTROLLED CURRENT SOURCE: $R_L = 2.2$ kΩ

| V_{in} | Calculated I_L | Measured I_L | |
		Multisim	Actual
0 V			
1 V			
2 V			
3 V			
4 V			
5 V			

TABLE 49-5. TROUBLESHOOTING

| Trouble | Estimated I_L | Measured I_L | |
		Multisim	Actual
R_3 shorted			
R_4 open			
Q_1 open			
Q_2 open			

TABLE 49-6. CRITICAL THINKING: $R =$ _____

	Multisim	% Error	Actual	% Error
I_L				

Questions for Experiment 49

1. In Table 49-1, the maximum load current is closest to: ( )
   (a) 1.5 mA; **(b)** 25 mA; **(c)** 50 mA; **(d)** 150 mA.
2. In Table 49-2, the load current for a 10 Ω resistor is closest to: ( )
   (a) 1.5 mA; **(b)** 25 mA; **(c)** 50 mA; **(d)** 150 mA.
3. The measured data of Tables 49-3 and 49-4 indicate that the load resistance is ( )
   driven by a(n):
   (a) voltage source; **(b)** current source; **(c)** transistor; **(d)** op amp.
4. The last measured entry of Table 49-4 indicates that the load voltage exceeds: ( )
   (a) V_{CC}; **(b)** $V_{CC} - v_{in}$; **(c)** V_{EE}; **(d)** $I_{out(max)}$.
5. As R_L increases in Fig. 49-3, the maximum input voltage: ( )
   (a) decreases; **(b)** increases; **(c)** stays the same; **(d)** equals zero.
6. Explain why the circuit of Fig. 49-3 cannot produce 5 mA when R_L is 2.2 kΩ.

TROUBLESHOOTING

7. Why does the load current decrease to zero when R_3 is shorted in Fig. 49-3?

CRITICAL THINKING

8. The circuit of Fig. 49-2 was built with a 2N3904 instead of a 2N3055. The circuit will not work with a 10 Ω load. What is the problem?

9. Optional. Instructor's question.

10. Optional. Instructor's question.

Active Low-Pass Filters

The low-pass filter is the prototype for filter analysis and design. Computer programs convert any filter design problem into an equivalent low-pass design problem. After the computer finds the low-pass components, it converts them into equivalent components for other responses. The low-pass filter is a basic introduction to filters.

In this experiment, first-order low-pass filters (one-pole), second-order low-pass filters (two-pole), and an optional fourth-order low-pass filter (four-pole) will be built and their frequency responses observed. To keep the wiring simple, Sallen-Key filters will be used. This class of filters produces excellent results in a laboratory when part selection is limited. The more complicated filters require too many nonstandard resistance and capacitance values.

Besides measuring the cutoff frequency and rolloff rate of Butterworth, Bessel, and Chebyshev filters, the step response of these filters will be observed. Recall that the step response is important in digital communications. The Bessel filter has the slowest rolloff but the best step response. Of the three, the Chebyshev filter has the fastest rolloff but the worst step response. The Butterworth filter is a good compromise, giving good rolloff and good step response.

Because active filters may use nonstandard capacitances, it is sometimes necessary to connect two standard capacitors in parallel to get a nonstandard value. For instance, to get 4.4 nF, use two 2.2-nF capacitors in parallel. The same idea applies to any circuit that does not have the required capacitance value: Put capacitors in parallel as needed to get as close as possible to the theoretical value.

GOOD TO KNOW

A first-order filter has a rolloff rate of 20 dB per decade. A second-order filter has a rolloff rate of 40 dB per decade and so forth.

Required Reading

Chapter 19 (Secs. 19-1 to 19-6) of *Electronic Principles,* 8th ed.

Equipment

- 1 function generator
- 2 power supplies: ±15 V
- 1 DMM (digital multimeter)
- 1 op amp: 741C
- 12 ½-W resistors: 1 kΩ, 4.7 kΩ, 8.2 kΩ, two 10 kΩ, 14 kΩ, two 33 kΩ, two 39 kΩ, two 47 kΩ
- 8 capacitors: three 2.2 nF, 3 nF, two 4.7 nF, 10 nF, 68 nF
- 1 oscilloscope

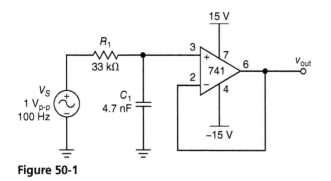

Figure 50-1

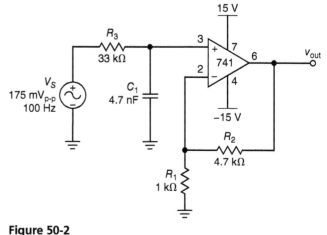

Figure 50-2

Procedure

BUTTERWORTH FIRST-ORDER LP FILTER

1. In Fig. 50-1, the calculated cutoff frequency and the attenuation one decade above f_C are:
$$f_C = 1026 \text{ Hz}$$
$$A_{V(dB)} = 20 \text{ dB}$$

2. Build the circuit shown in Fig. 50-1.

3. Use channel 1 of the oscilloscope to look at the input signal (left end of 33 kΩ), and use channel 2 for the output signal. Adjust the input signal to 1 V_{p-p} and 100 Hz.

4. Since the filter has unity gain in the passband, the output voltage should be equal to the input voltage.

5. Find the cutoff frequency by increasing the generator frequency until the output signal is down 3 dB from the input signal. (This occurs when the output signal equals 0.707 V_{p-p} when the input equals 1 V_{p-p}.)

6. Increase the generator frequency to 10 times the cutoff frequency. Measure the output voltage. Record the cutoff frequency and the attenuation one decade above the cutoff frequency:

$$f_C = \underline{\hspace{2cm}}.$$

$$A_{V(dB)} = \underline{\hspace{2cm}}.$$

7. The data in the preceding step should be reasonably close to the theoretical values of Step 1.

8. Use a square-wave input. Adjust the input signal to get 1 V_{p-p} and 100 Hz. Use a time base of 1 ms/DIV. Channel 1 will display the input square wave, and channel 2 will show the step response.

9. Since this is a Butterworth first-order filter, there is no overshoot or ringing on either the positive or the negative step. Section 14-9 in the textbook discusses the risetime-bandwidth relationship of any first-order response.

BUTTERWORTH FIRST-ORDER LP FILTER WITH GAIN

10. In Fig. 50-2, calculate the cutoff frequency and attenuation one decade above cutoff:

$$f_C = \underline{\hspace{2cm}}.$$

$$A_{V(dB)} = \underline{\hspace{2cm}}.$$

11. Because the voltage gain is ideally 5.7, an input signal of approximately 175 mV_{p-p} will produce an output signal of 1 V_{p-p}. Build the circuit and adjust the function generator as needed to get an output of 1 V_{p-p} and 100 Hz.

12. Measure the cutoff frequency and attenuation one decade above the cutoff frequency. The values you record should agree with those calculated in Step 10.

$$f_C = \underline{\hspace{2cm}}.$$

$$A_{V(dB)} = \underline{\hspace{2cm}}.$$

BUTTERWORTH FIRST-ORDER INVERTING LP FILTER

13. Theoretically, the filter of Fig. 50-3 has the same cutoff frequency and rolloff as the two preceding filters. Build the circuit and verify the cutoff frequency and attenuation one decade above cutoff:

$$f_C = \underline{\hspace{2cm}}.$$

$$A_{V(dB)} = \underline{\hspace{2cm}}.$$

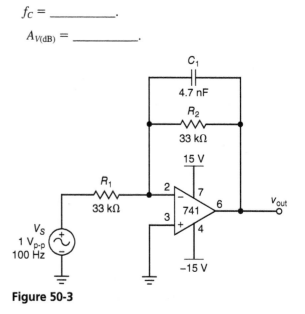

Figure 50-3

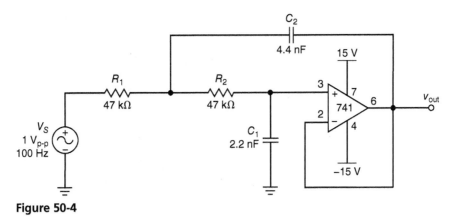

Figure 50-4

BUTTERWORTH SECOND-ORDER LP FILTER

14. In Fig. 50-4, calculate the cutoff frequency, attenuation one decade above cutoff, and the Q.

$f_C =$ _____.

$A_{V(dB)} =$ _____.

$Q =$ _____.

15. Based on the data in Step 14, state why the filter has a Butterworth response:

16. Build the circuit using two 2.2-nF in parallel for the 4.4 nF. Find the cutoff frequency and measure the attenuation one decade above cutoff:

$f_C =$ _____.

$A_{V(dB)} =$ _____.

17. Use a square-wave input as in Step 8 to look at the output step response. This time, some overshoot will be visible, as discussed in the textbook.

CHEBYSHEV SECOND-ORDER LP FILTER

18. In Fig. 50-5, calculate the cutoff frequency, attenuation one decade above cutoff, and the Q.

$f_C =$ _____.

$A_{V(dB)} =$ _____.

$Q =$ _____.

19. Based on the data in Step 18 and Fig. 19-25 in the textbook, state why the filter has a Chebyshev response:

20. Build the circuit. (*Note:* If a 68-nF capacitor is not available, use 47 nF in parallel with 22 nF.) Find the cutoff frequency and measure the attenuation one decade above cutoff:

$f_C =$ _____.

$A_{V(dB)} =$ _____.

21. Use a square-wave input, as was done earlier, to look at the output step response. This time, even more

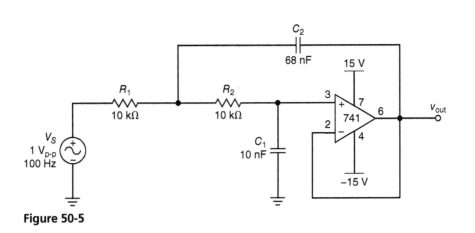

Figure 50-5

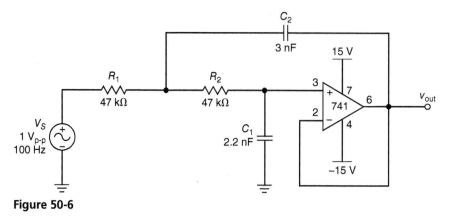

Figure 50-6

overshoot and ringing will be visible than with the second-order Butterworth filter.

22. Sketch the output waveform superimposed on the input waveform to show the overshoot and ringing of the Chebyshev filter:

BESSEL SECOND-ORDER LP FILTER

23. In Fig. 50-6, calculate the cutoff frequency, attenuation one decade above cutoff, and the Q.

$f_C =$ _____.

$A_{V(dB)} =$ _____.

$Q =$ _____.

24. Based on the data in Step 23, state why the filter has a Bessel response:

25. Build the circuit. Find the cutoff frequency and measure the attenuation one decade above cutoff:

$f_C =$ _____.

$A_{V(dB)} =$ _____.

26. Use a square-wave input, as was done earlier, to look at the output step response. This time, no overshoot will be visible because the Bessel has the best step response of any filter. (If there is any overshoot, the tolerance of the parts may account for it.)

27. Sketch the output waveform superimposed on the input waveform to show the step response of the Bessel filter:

BUTTERWORTH EQUAL-COMPONENT SECOND-ORDER LP FILTER

28. In Fig. 50-7, calculate the cutoff frequency, attenuation one decade above cutoff, and the Q.

$f_C =$ _____.

$A_{V(dB)} =$ _____.

$Q =$ _____.

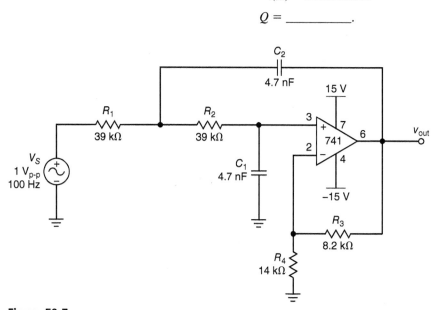

Figure 50-7

29. Based on the data in Step 28, state why the filter has a Butterworth response:

30. Build the circuit. Find the cutoff frequency and measure the attenuation one decade above cutoff:

$f_C =$ _____.

$A_{V(dB)} =$ _____.

31. Write a short essay on the basic concepts used to describe active low-pass filters. Be sure to support any conclusions drawn with factual data.

ADDITIONAL WORK (OPTIONAL)— BUTTERWORTH FOURTH-ORDER LP FILTER

32. Fig. 50-8 has a fourth-order Butterworth response. The two stages ideally have the same cutoff frequency, but the Qs are staggered to maintain a maximally flat response.

33. Calculate the cutoff frequency and Q of each stage:

$f_{C1} =$ _____.

$Q_1 =$ _____.

$f_{C2} =$ _____.

$Q_2 =$ _____.

34. Build the circuit. Find the cutoff frequency and measure the attenuation one decade above cutoff:

$f_C =$ _____.

$A_{V(dB)} =$ _____.

35. Use a square-wave input, as was done earlier, to look at the output step response. Sketch the step response:

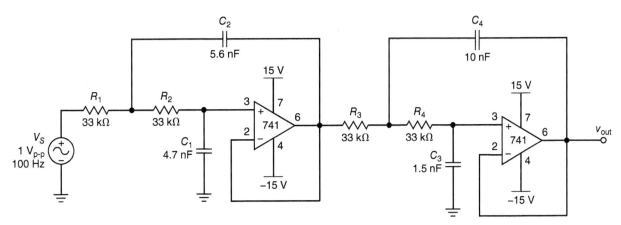

Figure 50-8

Experiment 50

Lab Partner(s) _____

PARTS USED Nominal Value	Measured Value	PARTS USED Nominal Value	Measured Value	PARTS USED Nominal Value	Measured Value
1 kΩ		33 kΩ		2.2 nF	
4.7 kΩ		39 kΩ		3 nF	
8.2 kΩ		39 kΩ		4.7 nF	
10 kΩ		47 kΩ		4.7 nF	
10 kΩ		47 kΩ		10 nF	
14 kΩ		2.2 nF		68 nF	
33 kΩ		2.2 nF			

CALCULATIONS

(Be sure to label all the answers so they reflect the procedure step they refer to.)

Active Butterworth Filters

The preceding experiment focused on Butterworth, Chebyshev, and Bessel responses. In this experiment, the focus is on the Butterworth response of low-pass, high-pass, wide bandpass, and narrow bandpass filters. The low-pass and high-pass filters will be Sallen-Key designs with a two-pole Butterworth response. The wide bandpass filter will be a cascade of a one-pole low-pass filter and a one-pole high-pass filter. The narrow bandpass filter will be a multiple-feedback filter with a two-pole Butterworth response.

Incidentally, the use of nanofarads has become widespread in active-filter design. Although many catalogs of electronic parts may list capacitance values like 0.001, 0.01, and 0.1 μF, most active-filter designers prefer using 1, 10, and 100 nF. Because of this, schematic diagrams of active filters typically use nanofarads when the capacitance is between 1 and 999 nF (equivalent to 0.001 and 0.999 μF).

GOOD TO KNOW

A wide bandpass filter has a Q less than 1, and a narrow bandpass filter has a Q greater than 1.

$$BW = f_2 - f_1$$
$$f_0 = \sqrt{f_1 f_2}$$
$$Q = \frac{f_0}{BW}$$

Required Reading

Chapter 19 (Secs. 19-1 to 19-10) of *Electronic Principles*, 8th ed.

Equipment

1 function generator
2 power supplies: ±15 V
1 DMM (digital multimeter)
2 op amps: 741C
9 ½-W resistors: 120 Ω, three 12 kΩ, two 22 kΩ, 33 kΩ, 39 kΩ, and 47 kΩ
6 capacitors: 1 nF, three 4.7 nF, two 33 nF
1 oscilloscope

Procedure

BUTTERWORTH SECOND-ORDER LOW-PASS FILTER

1. Calculate the cutoff frequency in Fig. 51-1. Record the value here:

   $f_c = $ _____.

2. What are the approximate attenuations one decade above and one decade below the cutoff frequency? Record the values here:

   Above = _____   Below = _____.

3. Measure and record the resistor and capacitor values. Build the circuit shown in Fig. 51-1.

4. Use channel 1 of the oscilloscope to look at the input signal and use channel 2 for the output signal. Adjust the input signal to 1 V_{p-p} and 100 Hz.

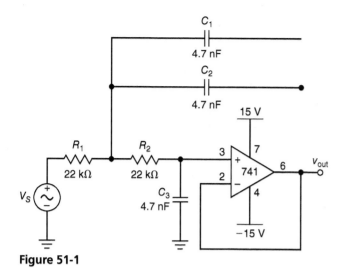

Figure 51-1

BUTTERWORTH SECOND-ORDER HIGH-PASS FILTER

11. Calculate the cutoff frequency in Fig. 51-2. Record the value here:

$$f_c = \underline{\qquad}.$$

12. What are the approximate attenuations one decade below and one decade above the cutoff frequency? Record the values here:

Below = _____ Above = _____.

13. Build the circuit shown in Fig. 51-2.
14. Use channel 1 of the oscilloscope to look at the input signal, and use channel 2 for the output signal. Adjust the input signal to 1 V_{p-p} and 20 kHz.
15. Since the filter has unity gain in the passband, the output voltage should be equal to the input voltage.
16. Find the cutoff frequency by decreasing the generator frequency until the output signal is down 3 dB from the input signal. Record the cutoff frequency here:

$$f_c = \underline{\qquad}.$$

17. Decrease the generator frequency to one-tenth of the cutoff frequency. Measure and record the peak-to-peak output voltage:

$$v_{out} = \underline{\qquad}.$$

18. Increase the generator frequency to 10 times the cutoff frequency. Measure and record the peak-to-peak output voltage:

$$v_{out} = \underline{\qquad}.$$

19. The data in the preceding step should be reasonably close to the theoretical values of Steps 11 and 12.
20. Measure and record the output voltage for each of the input frequencies shown in Table 51-2.

5. Since the filter has unity gain in the passband, the output voltage should be equal to the input voltage.
6. Find the cutoff frequency by increasing the generator frequency until the output signal is down 3 dB from the input signal. (This means that the output signal equals 0.707 V_{p-p} when the input equals 1 V_{p-p}.) Record the cutoff frequency here:

$$f_c = \underline{\qquad}.$$

7. Increase the generator frequency to 10 times the cutoff frequency. Measure and record the peak-to-peak output voltage:

$$v_{out} = \underline{\qquad}.$$

8. Decrease the generator frequency to one-tenth of the cutoff frequency. Measure and record the peak-to-peak output voltage:

$$v_{out} = \underline{\qquad}.$$

9. The data in the preceding step should be reasonably close to the theoretical values of Steps 1 and 2.
10. Measure and record the output voltage for each of the input frequencies shown in Table 51-1.

BUTTERWORTH WIDE BANDPASS FILTER

21. Calculate the lower and upper cutoff frequencies in Fig. 51-3. Record the values here:

$$f_1 = \underline{\qquad} \qquad f_2 = \underline{\qquad}.$$

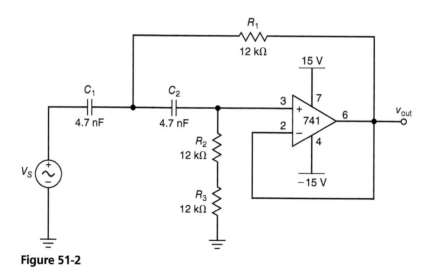

Figure 51-2

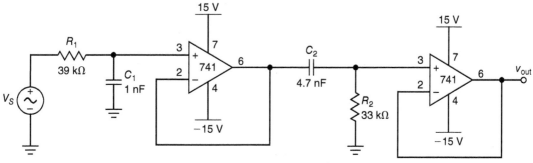

Figure 51-3

22. Calculate the geometric center frequency and record the value here:

$f_0 =$ _____.

23. What are the approximate attenuations one decade below the lower cutoff frequency and one decade above the upper cutoff frequency? Record the values here:

Below = _____ Above = _____.

24. Build the circuit shown in Fig. 51-3.
25. Use channel 1 of the oscilloscope to look at the input signal, and use channel 2 for the output signal.
26. Adjust the input signal to 1 V_{p-p} and 2 kHz.
27. Find the lower cutoff frequency by decreasing the generator frequency until the output signal is down 3 dB from the input signal. Record the lower cutoff frequency here:

$f_1 =$ _____.

28. Find the upper cutoff frequency by increasing the generator frequency until the output signal is down 3 dB from the input signal. Record the upper frequency here:

$f_2 =$ _____.

29. The two cutoff frequencies of Steps 27 and 28 should be reasonably close to the theoretical values of Step 21.
30. Measure and record the output voltage for each of the input frequencies shown in Table 51-3.

BUTTERWORTH NARROWBAND BANDPASS FILTER

31. Calculate the center frequency and bandwidth in Fig. 51-4. Record the values here:

$f_0 =$ _____ $BW =$ _____.

32. Build the circuit shown in Fig. 51-4.
33. Use channel 1 of the oscilloscope to look at the input signal, and use channel 2 for the output signal.
34. Adjust the input signal to 1 V_{p-p} and 2 kHz.
35. Find the center frequency by varying the generator frequency until the output signal is maximum. Record the center frequency here:

$f_0 =$ _____.

36. Find the lower cutoff frequency by decreasing the generator frequency until the output signal is down 3 dB. Record the lower frequency here:

$f_1 =$ _____.

37. Find the upper cutoff frequency by increasing the generator frequency until the output signal is down 3 dB. Record the upper frequency here:

$f_2 =$ _____.

38. What is the bandwidth? Record the value here:

$BW =$ _____.

39. Measure and record the output voltage for each of the input frequencies shown in Table 51-4.

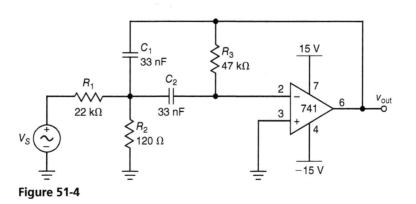

Figure 51-4

Experiment 51

Lab Partner(s) _____

PARTS USED Nominal Value	Measured Value	PARTS USED Nominal Value	Measured Value	PARTS USED Nominal Value	Measured Value
120 Ω		22 kΩ		4.7 nF	
12 kΩ		33 kΩ		4.7 nF	
12 kΩ		39 kΩ		4.7 nF	
12 kΩ		47 kΩ		33 nF	
22 kΩ		1 nF		33 nF	

TABLE 51-1. TWO-POLE LOW-PASS RESPONSE

Frequency	Multisim Output Voltage	Actual Output Voltage
100 Hz		
200 Hz		
400 Hz		
1 kHz		
2 kHz		
4 kHz		
10 kHz		
20 kHz		
40 kHz		

TABLE 51-2. TWO-POLE HIGH-PASS RESPONSE

Frequency	Multisim Output Voltage	Actual Output Voltage
100 Hz		
200 Hz		
400 Hz		
1 kHz		
2 kHz		
4 kHz		
10 kHz		
20 kHz		
40 kHz		

297

TABLE 51-3. WIDE BANDPASS RESPONSE

Frequency	Multisim Output Voltage	Actual Output Voltage
100 Hz		
200 Hz		
400 Hz		
1 kHz		
2 kHz		
4 kHz		
10 kHz		
20 kHz		
40 kHz		

TABLE 51-4. NARROW BANDPASS RESPONSE

Frequency	Multisim Output Voltage	Actual Output Voltage
100 Hz		
200 Hz		
400 Hz		
1 kHz		
2 kHz		
4 kHz		
10 kHz		
20 kHz		
40 kHz		

Questions for Experiment 51

1. The circuit of Fig. 51-1 has a cutoff frequency closest to: ( )
   (a) 100 Hz;   (b) 1 kHz;   (c) 2 kHz;   (d) 10 kHz.
2. The circuit of Fig. 51-1 has a Butterworth response because it: ( )
   (a) is a low-pass filter;   (b) has equal resistors;   (c) has a Q of 0.707;
   (d) has a cutoff frequency.
3. The circuit of Fig. 51-2 has what kind of response? ( )
   (a) low-pass;   (b) high-pass;   (c) wide bandpass;   (d) narrow bandpass.
4. The circuit of Fig. 51-2 has a cutoff frequency closest to: ( )
   (a) 100 Hz;   (b) 1 kHz;   (c) 2 kHz;   (d) 10 kHz.
5. The wide bandpass filter of Fig. 51-3 has a lower cutoff frequency closest to: ( )
   (a) 100 Hz;   (b) 1 kHz;   (c) 2 kHz;   (d) 4 kHz.
6. The wide bandpass filter of Fig. 51-3 has an upper cutoff frequency closest to: ( )
   (a) 100 Hz;   (b) 1 kHz;   (c) 2 kHz;   (d) 4 kHz.
7. The narrow bandpass filter of Fig. 51-4 has a center frequency closest to: ( )
   (a) 100 Hz;   (b) 1 kHz;   (c) 2 kHz;   (d) 4 kHz.

8. The narrow bandpass filter of Fig. 51-4 has a Q closest to: ( )
    (a) 1;   **(b)** 2;   **(c)** 4;   **(d)** 10.

9. Optional. Instructor's question.

10. Optional. Instructor's question.

Active Diode Circuits and Comparators

With the help of an op amp, the effect of the diode knee voltage can be reduced. The effective knee voltage is reduced by the open-loop gain of the op amp. For a typical 741C, this means the equivalent knee voltage is only 7 μV. This allows the building of circuits that will rectify, peak-detect, limit, and clamp low-level signals.

A comparator is a circuit that can indicate when the input voltage exceeds a specific limit. With a zero-crossing detector, the trip point is zero. With a limit detector, the trip point is either a positive or negative voltage.

In this experiment, a variety of active diode circuits as well as a zero-crossing detector and a limit detector will be built.

GOOD TO KNOW

Function generators often produce a noisy signal at very small amplitudes. A voltage divider can be used when a small input signal is required.

Required Reading

Chapter 20 (Secs. 20-1 to 20-9) of *Electronic Principles*, 8th ed.

Equipment

1   function generator
2   power supplies: ± 15 V
6   ½-W resistors: 100 Ω, 1 kΩ, 2.2 kΩ, two 10 kΩ, 100 kΩ
1   potentiometer: 1 kΩ
1   diode: 1N4148 or 1N914
2   LEDs: L53RD and L53GD (or similar red and green LEDs)
1   op amp: 741C
3   capacitors: two 0.47 μF, 100 μF (rated at least 15 V)
1   DMM (digital multimeter)
1   oscilloscope

Procedure

HALF-WAVE RECTIFIER

1. Measure and record the resistor and capacitor values. Build the circuit shown in Fig. 52-1.

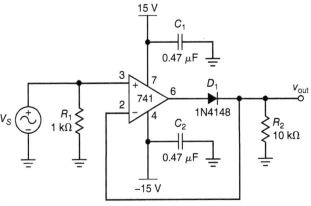

Figure 52-1

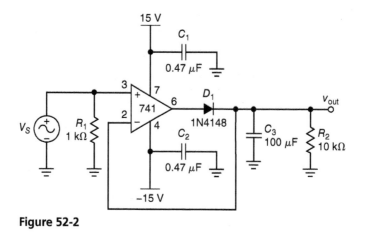

Figure 52-2

2. Connect the oscilloscope (dc input) across the 10 kΩ load resistor. Set the generator to 100 Hz and adjust the level to get a peak output of 1 V on the oscilloscope. (A half-wave signal should be visible on the oscilloscope.)

3. Measure the peak value of the input sine wave. Record the input and output peak voltages in Table 52-1.

4. Adjust the signal level to get a half-wave output with a peak value of 100 mV. Then measure the peak input voltage. Record the input and output peak voltages in Table 52-1.

5. Reverse the polarity of the diode. The output voltage should be a negative half-wave signal.

PEAK DETECTOR

6. Connect a 100 μF capacitor across the load to get the circuit of Fig. 52-2.

7. Adjust the generator to get an input peak value of 1 V. Measure the dc output voltage. Record the peak input voltage and the dc output voltage in Table 52-2.

8. Readjust the generator to get an input peak value of 100 mV. Measure the dc output voltage. Record the peak input voltage and the dc output voltage in Table 52-2.

9. Reverse the polarity of the diode and the capacitor. The output voltage should be a negative dc voltage.

LIMITER

10. Build the circuit shown in Fig. 52-3.

11. Adjust the generator to produce a peak-to-peak value of 1 V at the left end of the 2.2 kΩ resistor.

12. Observe the output signal while turning the potentiometer through its entire range.

13. Adjust the generator to produce a peak-to-peak output of 100 mV at the left end of the 2.2 kΩ resistor. Then repeat Step 12.

14. Reverse the polarity of the diode and repeat Step 12 for a peak-to-peak output of 1 V.

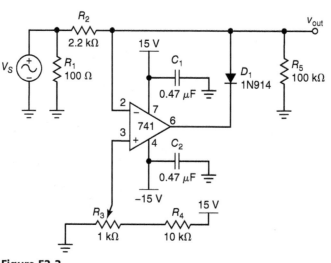

Figure 52-3

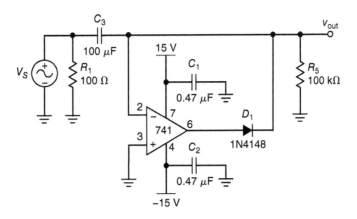

Figure 52-4

DC CLAMPER

15. Build the circuit shown in Fig. 52-4.
16. Adjust the input to 1 V_{p-p}.
17. Observe the output. It should be a positively clamped signal.
18. Reduce the input to 100 mV_{p-p} and repeat Step 17.
19. Reverse the polarity of the diode. The output should now be negatively clamped.

ZERO-CROSSING DETECTOR

20. Build the zero-crossing detector of Fig. 52-5. (*Note:* The I_{max} out of the 741C is approximately 25 mA, so the LED current is limited to 25 mA. If an op amp has an I_{max} greater than 50 mA, current-limiting resistors will be required because most LEDs cannot handle more than 50 mA.)
21. Vary the potentiometer and notice what the LEDs do.
22. Use the dc-coupled input of the oscilloscope to look at the input voltage to pin 3. Adjust the potentiometer

to get +100 mV at the input. Record the input voltage and the color of the LED that is on (Table 52-3).
23. Adjust the potentiometer to get an input of −100 mV. Record the input voltage and the color of the LED that is on.

LIMIT DETECTOR

24. In Fig. 52-6, calculate the trip point of the limit detector. Record the answer in Table 52-4.
25. Build the circuit. Adjust the input voltage until the approximate trip point is found. Record the trip point.

TROUBLESHOOTING

26. For each set of symptoms listed in Table 52-5, try to figure out what trouble could produce the symptoms in Fig. 52-6. Insert the trouble and verify that it produces the symptoms. Record each trouble in Table 52-5.

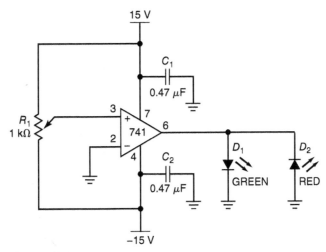

Figure 52-5

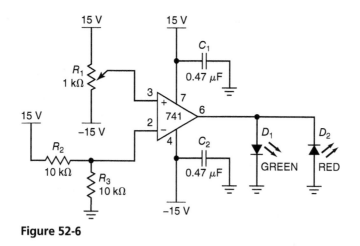

Figure 52-6

CRITICAL THINKING

27. Select a value of R_3 to get a trip point of approximately +5 V. Connect the circuit with your design value and measure the trip point. Record the quantities listed in Table 52-6.

APPLICATION (OPTIONAL)

28. Figure 52-7 shows a window comparator similar to Fig. 20-22 in the textbook. A circuit like this can be used for go/no-go testing. If the voltage being tested is outside its normal range, an LED can be lit or a buzzer sounded.
29. Build the circuit shown in Fig. 52-7.
30. Vary the potentiometer and notice that the red LED goes on and off.
31. With the LED on, vary the potentiometer slowly until the LED goes off. Record the input voltage:

   _____.

32. With the LED off, vary the potentiometer slowly until it goes on. Record the input voltage:

   _____.

33. Replace the LED with a buzzer MCM Part #: 28-740.
34. Repeat Steps 31 and 32 for the buzzer:

   $V_1 =$ _____.

   $V_2 =$ _____.

35. What was learned about a window comparator?

36. How could the window comparator of Fig. 52-7 be modified to monitor a circuit to ensure its dc output stays between 2.5 and 3.5 V?

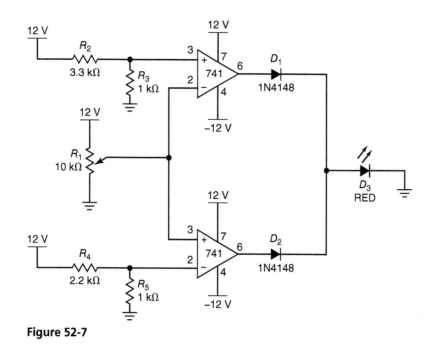

Figure 52-7

Experiment 52

Lab Partner(s) _____

PARTS USED Nominal Value	Measured Value	PARTS USED Nominal Value	Measured Value	PARTS USED Nominal Value	Measured Value
100 Ω		10 kΩ		0.47 μF	
1 kΩ		10 kΩ		0.47 μF	
2.2 kΩ		100 kΩ		100 μF	

TABLE 52-1. ACTIVE HALF-WAVE RECTIFIER

	Multisim		Actual	
	v_{in}	v_{out}	v_{in}	v_{out}
Step 3				
Step 4				

TABLE 52-2. ACTIVE PEAK DETECTOR

	Multisim		Actual	
	v_{in}	v_{out}	v_{in}	v_{out}
Step 7				
Step 8				

TABLE 52-3. ZERO-CROSSING DETECTOR

	Multisim		Actual	
	v_{in}	Color	v_{in}	Color
Step 22				
Step 23				

TABLE 52-4. LIMIT DETECTOR

	Calculated	Measured	
		Multisim	Actual
Trip point			

TABLE 52-5. TROUBLESHOOTING

Symptoms	Trouble
1. Red LED always on	
2. Trip point equals zero	
3. Red LED goes on and off, but green LED is always off	
4. Neither LED comes on	

TABLE 52-6. CRITICAL THINKING: R_3 = _____

	Multisim	% Error	Actual	% Error
Trip point				

Questions for Experiment 52

1. The circuit of Fig. 52-1 is a: ( )
   (a) half-wave rectifier;   (b) full-wave rectifier;   (c) bridge rectifier;
   (d) none of the foregoing.

2. The dc output voltage of Fig. 52-2 is approximately equal to the: ( )
   (a) peak input voltage;   (b) positive supply voltage;   (c) rms input
   voltage;   (d) average input voltage.

3. The positive limiter of Fig. 52-3 can be adjusted to have a limiting level between ( )
   0 and approximately:
   (a) +10 V;   (b) +1.36 V;   (c) −5 V;   (d) +12 V.

4. The circuit of Fig. 52-4 clamps the signal: ( )
   (a) negatively;   (b) positively;   (c) at −5 V;   (d) at +3 V.

5. The limit detector of Fig. 52-6 has a trip point of approximately: ( )
   (a) 0;   (b) +5 V;   (c) +7.5 V;   (d) +10 V.

6. Explain how the limit detector of Fig. 52-6 works.

TROUBLESHOOTING

7. Name at least two troubles in Fig. 52-6 that would produce a trip point of zero.

8. If the capacitor of Fig. 52-2 opens, what symptoms will be present?

CRITICAL THINKING

9. How was the design value for R_3 arrived at?

10. Optional. Instructor's question.

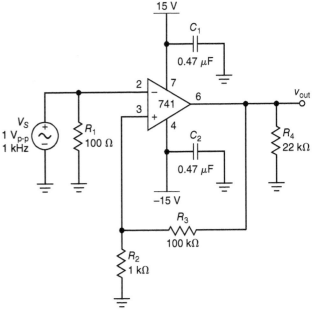

Figure 53-1

Waveshaping Circuits

Experiment **53**

By using positive feedback with a comparator, the Schmitt trigger can be constructed. It has hysteresis, which makes it less sensitive to noise. A Schmitt trigger is useful for waveshaping because it produces a square-wave output no matter what the shape of the input signal.

A relaxation oscillator is created by adding an *RC* circuit to a Schmitt trigger. This type of circuit generates a square-wave output without an external input signal. A circuit that generates square waves and triangular waves can be built by cascading a relaxation oscillator and an integrator.

GOOD TO KNOW

The oscillation frequency of the relaxation oscillator is based on the time it takes for the capacitor to charge and discharge.

Required Reading

Chapter 20 (Secs. 20-3 to 20-8) of *Electronic Principles,* 8th ed.

Equipment

1 function generator
2 power supplies: adjustable from 0 to 15 V
8 ½-W resistors: 100 Ω, 1 kΩ, two 2.2 kΩ, 10 kΩ, 18 kΩ, 22 kΩ, 100 kΩ
3 op amps: 318C, two 741C
6 capacitors: two 0.1 μF, four 0.47 μF
1 oscilloscope
1 frequency counter
1 DMM (digital multimeter)

Procedure

SCHMITT TRIGGER

1. In Fig. 53-1, what shape will the output signal have? Estimate the peak-to-peak output voltage. Record these answers in Table 53-1. Also calculate and record the trip points.

2. Measure and record the resistor and capacitor values. Build the circuit. Adjust the input voltage to 1 V_{p-p} at 1 kHz.
3. Observe the output with an oscilloscope. Record the approximate shape of the signal in Table 53-1. Also measure and record the peak-to-peak output voltage.
4. Observe the noninverting input voltage with the oscilloscope on dc input. Measure the positive peak and record this as the UTP. Measure the negative peak and record as the LTP.

EFFECT OF SLEW-RATE LIMITING

5. Increase the frequency to 20 kHz. The output should be approximately rectangular. (*Note:* Some overshoot or ringing may be present because of the high slew rate of a 318C, but the transitions between high and low should still appear almost vertical.)
6. Change the frequency back to 1 kHz. Replace the 318C with a 741C. The output should appear approximately rectangular.
7. Increase the frequency and notice how the slew rate of a 741C affects the vertical transitions.

RELAXATION OSCILLATOR AND INTEGRATOR

8. In Fig. 53-2, a relaxation oscillator drives an integrator. Calculate the frequency out of the relaxation oscillator. Record in Table 53-2.

9. Assume $+V_{sat}$ is $+14$ V and $-V_{sat}$ is -14 V. Calculate and record the peak-to-peak output voltage from the relaxation oscillator.
10. Calculate and record the peak-to-peak output voltage from the integrator.
11. Connect the circuit.
12. Observe the signal out of the relaxation oscillator. Measure its frequency and peak-to-peak value. Record this data.
13. Look at the signal at the inverting input of the relaxation oscillator. It should look like Fig. 20-22b in the textbook.
14. Look at the signal out of the integrator. Measure and record its peak-to-peak value.

TROUBLESHOOTING

15. For each trouble listed in Table 53-3, calculate the output frequency and the peak-to-peak output voltage in Fig. 53-2. Record the answers.
16. Insert each trouble into the circuit. Measure and record the output frequency and peak-to-peak voltage.

CRITICAL THINKING

17. Select a value of R_1 in Fig. 53-2 to get a frequency of approximately 1 kHz.
18. Build the circuit with the design value of R_1. Measure the frequency. Record the value of R_1 and frequency in Table 53-4.

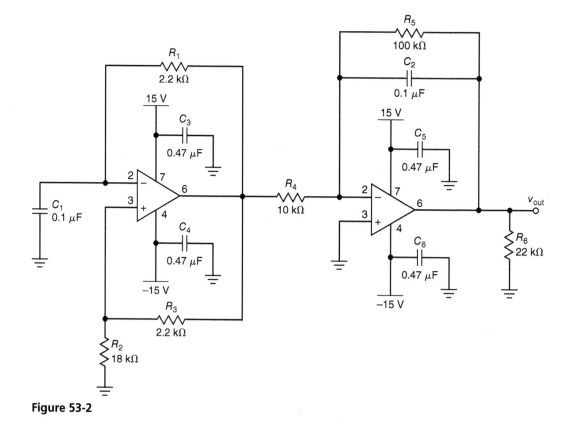

Figure 53-2

Experiment 53

Lab Partner(s) _____

PARTS USED Nominal Value	Measured Value	PARTS USED Nominal Value	Measured Value	PARTS USED Nominal Value	Measured Value
100 Ω		18 kΩ		0.47 μF	
1 kΩ		22 kΩ		0.47 μF	
2.2 kΩ		100 kΩ		0.47 μF	
2.2 kΩ		0.1 μF		0.47 μF	
10 kΩ		0.1 μF			

TABLE 53-1. SCHMITT TRIGGER

	Shape	MPP	UTP	LTP
Calculated				
Multisim				
Actual				

TABLE 53-2. RELAXATION OSCILLATOR AND INTEGRATOR

Calculated			Measured					
			Multisim			Actual		
f	$v_{out(1)}$	$v_{out(2)}$	f	$v_{out(1)}$	$v_{out(2)}$	f	$v_{out(1)}$	$v_{out(2)}$

TABLE 53-3. TROUBLESHOOTING

			Measured			
	Calculated		Multisim		Actual	
Trouble	f	v_{out}	f	v_{out}	f	v_{out}
R_1 is 22 kΩ						
R_2 is 1.8 kΩ						
R_4 is 22 kΩ						

TABLE 53-4. CRITICAL THINKING: $R_1 = $ _____

	Multisim	% Error	Actual	% Error
f				

Questions for Experiment 53

1. The square wave out of the Schmitt trigger (Fig. 53-1) has a peak-to-peak value that is closest to: ( )
    (a) 5 V;   (b) 10 V;   (c) 20 V;   (d) 30 V.
2. The UTP of the Schmitt trigger was approximately: ( )
    (a) −0.1 V;   (b) +0.1 V;   (c) −10 V;   (d) +10 V.
3. The relaxation oscillator has a calculated frequency in Table 53-2 of: ( )
    (a) 345 Hz;   (b) 456 Hz;   (c) 796 Hz;   (d) 1.27 kHz.
4. The triangular output of the integrator in Fig. 53-2 has a calculated peak-to-peak value of approximately: ( )
    (a) 0.1 V;   (b) 8.79 V;   (c) 12.3 V;   (d) 15 V.
5. The waveform at the inverting input of the relaxation oscillator (Fig. 53-2) appears: ( )
    (a) square;   (b) triangular;   (c) exponential;   (d) sinusoidal.
6. Explain how a Schmitt trigger like Fig. 53-1 works.

7. Explain how a relaxation oscillator like Fig. 53-2 works.

TROUBLESHOOTING

8. What are the symptoms in Fig. 53-2 when R_1 is 22 kΩ instead of 2.2 kΩ? Why do these changes occur?

CRITICAL THINKING

9. Why is it better to use a 318C instead of a 741C in a Schmitt trigger?

10. Optional. Instructor's question.

The Wien-Bridge Oscillator

The Wien-bridge oscillator is the standard oscillator circuit for low to moderate frequencies in the range of 5 Hz to about 1 MHz. The oscillation frequency is equal to $1/2\pi RC$. Typically, a tungsten lamp is used to reduce the loop gain $A_V B$ to unity. It is also possible to use diodes, zener diodes, and JFETs as nonlinear elements to reduce the loop gain to unity. In this experiment, a Wien-bridge oscillator will be built and analyzed.

GOOD TO KNOW

The frequency of the output signal can also be measured with the oscilloscope. Recall that the frequency is equal to the reciprocal of the period of the waveform: $f = 1/T$.

Required Reading

Chapter 21 (Secs. 21-1 and 21-2) of *Electronic Principles,* 8th ed.

Equipment

2 power supplies: ± 15 V
9 ½-W resistors: two 1 kΩ, two 2.2 kΩ, two 4.7 kΩ, 8.2 kΩ, two 10 kΩ
3 diodes: 1N4148 or 1N914
1 LED: L53RD (or similar red LED)
1 op amp: 741C
1 potentiometer: 5 kΩ
4 capacitors: two 0.01 μF, two 0.47 μF
1 oscilloscope
1 frequency counter
1 DMM (digital multimeter)

Procedure

OSCILLATOR

1. In Fig. 54-1, calculate and record the frequency of oscillation for each value of R in Table 54-1. Also, calculate and record the MPP values. (*Note:* The LED is used to indicate when the circuit is oscillating. The 1N914 across the LED protects it during reverse bias because the LED has a breakdown voltage of only 3 V.)

2. Measure and record the resistor and capacitor values. Build the circuit shown in Fig. 54-1 with an R of 10 kΩ. While observing the output with an oscilloscope, adjust R_3 to get as large an unclipped output as possible.

3. Measure the frequency. Measure the peak-to-peak output voltage. Record both quantities in Table 54-1.

4. Repeat Steps 2 and 3 for the other values of R.

TROUBLESHOOTING

5. Insert each trouble listed in Table 54-2. Determine what effect the trouble has on the output signal. Record the symptoms in Table 54-2. (Examples of entries are "no output," "heavily clipped output," and "small distorted output.")

CRITICAL THINKING

6. Select a value of R (nearest standard value) to get an oscillation frequency of approximately 2.25 kHz.

7. Build the circuit with the selected value of R. Measure the frequency. Record the quantities in Table 54-3.

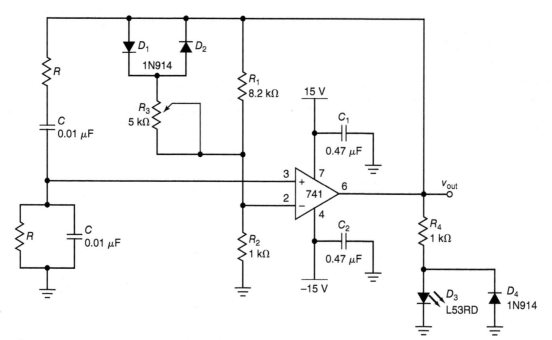

Figure 54-1

Experiment 54

Lab Partner(s) _____

PARTS USED Nominal Value	Measured Value	PARTS USED Nominal Value	Measured Value	PARTS USED Nominal Value	Measured Value
1 kΩ		4.7 kΩ		0.01 μF	
1 kΩ		8.2 kΩ		0.01 μF	
2.2 kΩ		10 kΩ		0.47 μF	
2.2 kΩ		10 kΩ		0.47 μF	
4.7 kΩ					

TABLE 54-1. OSCILLATOR

		Calculated		Measured			
				Multisim		Actual	
R		f	MPP	f	MPP	f	MPP
10 kΩ							
4.7 kΩ							
2.2 kΩ							

TABLE 54-2. TROUBLESHOOTING

	Symptoms	
Trouble	Multisim	Actual
R_1 shorted		
R_1 open		
R_2 shorted		
R_2 open		
R_3 shorted		
R_3 open		

TABLE 54-3. CRITICAL THINKING: $R =$ _____

	Multisim	% Error	Actual	% Error
f				

Questions for Experiment 54

1. The data of Table 54-1 indicate that an increase in resistance produces which of   ( )
   the following changes in oscillation frequency:
   (a) decrease;   (b) increase;   (c) no change.
2. The MPP value of the circuit was closest to: ( )
   (a) 0.7 V;   (b) 1.4 V;   (c) 15 V;   (d) 27 V.

3. The component that is involved in reducing loop gain to unity is the: ( )
   (a) 741C;   (b) LED;   (c) 1N4148;   (d) 0.01 μF.
4. The peak LED current is closest to: ( )
   (a) 8.59 mA;   (b) 17.1 mA;   (c) 19.1 mA;   (d) 27 mA.
5. The 1N914 protects the LED from excessive reverse voltage because the 1N914: ( )
   (a) breaks down first;   (b) conducts when the reverse voltage exceeds -0.7 V;
   (c) goes into reverse bias when the LED is on;   (d) has a higher power dissipation
   than the LED.
6. Briefly explain how a Wien-bridge oscillator works.

TROUBLESHOOTING

7. Explain why there is no output when R_1 is shorted.

8. Explain why the output is heavily clipped when R_3 is open.

CRITICAL THINKING

9. How can the Wien-bridge oscillator be designed so that it is tunable to different frequencies?

10. Optional. Instructor's question.

The LC Oscillator

For oscillation frequencies between approximately 1 and 500 MHz, the *LC* oscillator is used instead of a Wien-bridge oscillator. This type of oscillator uses a resonant *LC* tank circuit to determine the frequency. The Colpitts oscillator is the most widely used *LC* oscillator because the feedback voltage is conveniently produced by a capacitive voltage divider rather than an inductive divider (Hartley). For an oscillator to start, the small-signal voltage gain must be greater than the reciprocal of the feedback fraction. In symbols, $A_V > 1/B$. As the oscillations increase, the value of A_V decreases until the loop gain A_VB equals unity.

GOOD TO KNOW

The frequency of the output signal can also be measured with the oscilloscope. Recall that the frequency is equal to the reciprocal of the period of the waveform: $f = 1/T$.

Required Reading

Chapter 21 (Sec. 21-4) of *Electronic Principles,* 8th ed.

Equipment

1   power supply: 15 V
2   ½-W resistors: 22 kΩ, 47 kΩ
1   inductor: 33 mH (or nearest value)
4   capacitors: 100 pF, 1000 pF, 0.1 μF, 0.47 μF
1   transistor: 2N3904
1   potentiometer: 50 kΩ
1   oscilloscope
1   frequency counter
1   DMM (digital multimeter)

Procedure

COLPITTS OSCILLATOR

1. In Fig. 55-1, neglect transistor and stray-wiring capacitance. Calculate the frequency of oscillation. Also calculate the peak-to-peak output voltage and the feedback fraction. Record the answers in Table 55-1. (*Note:* The 0.47 μF capacitor is a supply bypass

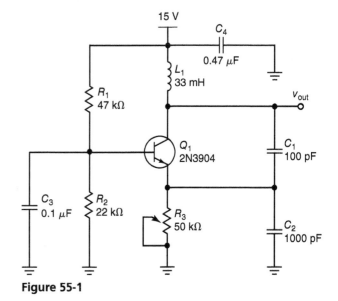

Figure 55-1

capacitor needed with some power supplies. This capacitor provides an ac ground at the upper end of L_1 and prevents the supply impedance from affecting the oscillation frequency and amplitude.)

2. Measure and record the resistor, inductor, and capacitor values. Build the circuit shown in Fig. 55-1.

3. Observe the emitter signal with channel 1 and the output signal with channel 2 of an oscilloscope.
4. Vary the 50 kΩ emitter resistor through its range until oscillations start. If oscillations do not start, try using a 100 kΩ potentiometer.
5. Adjust this resistor so that the output signal is just starting to clip.
6. Measure the frequency of oscillation and the peak-to-peak output voltage. Record the values.
7. Measure the peak-to-peak emitter signal. Calculate and record the feedback fraction B in Table 55-1.

TROUBLESHOOTING

8. Insert each trouble listed in Table 55-2. Record the output symptoms. (Examples of entries are "no output," "smaller output," and "higher frequency.")

CRITICAL THINKING

9. Ignore transistor and stray capacitance. Select new values for C_1 and C_2 to get an oscillation frequency that is 50 percent higher than the calculated frequency in Table 55-1.
10. Insert the selected values. Measure the oscillation frequency. Record all quantities in Table 55-3.

ADDITIONAL WORK (OPTIONAL)

11. As discussed in Experiment 31, a technician must be constantly aware of the loading effect of the oscilloscope probes. An inexpensive oscilloscope may add as much as 200 pF to a circuit on the X1 probe position. When using the X10 probe position, the input capacitance decreases to 20 pF. Therefore, use the X10 position whenever possible.
12. Probe capacitances will depend on the quality of the oscilloscope being used. Find out what the probe capacitances are for the X1 and X10 position of the oscilloscope. Record the values:

$C_{(X1)} = $ _____.

$C_{(X10)} = $ _____.

13. Replace the 33 mH inductor with a 330 μH inductor (or an inductor near this value). Since the resonant frequency is inversely proportional to the square root of inductance, the new resonant frequency will be approximately 10 times higher than before.
14. Observe the output with the X1 position of the probe. Measure the resonant frequency and record the value:

$f_r = $ _____ (X1).

15. Switch the probe to the X10 position. Increase the sensitivity by a factor of 10 to compensate for the probe attenuation of X10.
16. Repeat Step 14 and record the resonant frequency:

$f_r = $ _____ (X10).

17. The frequency in Step 16 should be slightly higher than the frequency in Step 14 because the probe will add less capacitance to the circuit during the measurement of resonant frequency.
18. When no probe is connected, is the resonant frequency higher, lower, or the same as the frequency in Step 16?

Answer = _____.

19. What was learned about the effect an oscilloscope probe has on a circuit under test conditions?

20. If all the wires were shortened as much as possible in the circuit, what effect would this have on the resonant frequency? Justify the answer given.

Experiment 55

Lab Partner(s) _____

PARTS USED Nominal Value	Measured Value	PARTS USED Nominal Value	Measured Value	PARTS USED Nominal Value	Measured Value
22 kΩ		100 pF		0.47 μF	
47 kΩ		1000 pF		33 mH	
50 kΩ Pot		0.1 μF			

TABLE 55-1. COLPITTS OSCILLATOR

Calculated			Measured					
			Multisim			Actual		
f	MPP	B	f	MPP	B	f	MPP	B

TABLE 55-2. TROUBLESHOOTING

Trouble	Output symptoms	
	Multisim	Actual
R_1 shorted		
R_1 open		
R_2 shorted		
R_2 open		
R_3 shorted		
R_3 open		
C_1 shorted		
C_1 open		
C_2 shorted		
C_2 open		
C_3 shorted		
C_3 open		

TABLE 55-3. CRITICAL THINKING: C_1 = _____; C_2 = _____

	Multisim	% Error	Actual	% Error
f				

Questions for Experiment 55

1. The calculated oscillation frequency of Fig. 55-1 is approximately: ( )
    (a) 90 kHz;   (b) 225 kHz;   (c) 445 kHz;   (d) 528 kHz.
2. The calculated feedback fraction of Fig. 55-1 is closest to: ( )
    (a) 0.091;   (b) 0.1;   (c) 1;   (d) 10.
3. For the oscillator to start, the minimum voltage gain is approximately: ( )
    (a) 1;   (b) 5;   (c) 11;   (d) 25.
4. The LC oscillator of Fig. 55-1 is an example of a: ( )
    (a) CB oscillator;   (b) CE oscillator;   (c) Wien-bridge oscillator;
    (d) twin-T oscillator.
5. The calculated peak-to-peak output voltage of Fig. 55-1 is approximately: ( )
    (a) 20 V;   (b) 25 V;   (c) 30 V;   (d) 40 V.
6. Briefly explain how an LC oscillator works.

7. In Fig. 55-1, what effect will transistor and stray-wiring capacitance have on the frequency of oscillation?

TROUBLESHOOTING

8. Why is there no output when R_1 is open?

CRITICAL THINKING

9. Explain why the actual frequency of oscillation will be less than the calculated frequency of oscillation.

10. Optional. Instructor's question.

Op-Amp Applications: Signal Generators

A function generator is a widely used instrument that produces three output shapes: sinusoidal, rectangular, and triangular. Fig. 56-1 shows a simple function generator that will be built in this experiment. The first stage is a Wien-bridge oscillator, the second stage is a Schmitt trigger, and the third stage is an integrator.

Each stage will be built and tested, starting with the Wien-bridge oscillator. After it is working properly, build and test the Schmitt trigger. Finally, connect the integrator, and the function generator will be complete. Point *A* will have a sinusoidal output, point *B* will have a rectangular output, and point *C* will have a triangular output.

GOOD TO KNOW

When troubleshooting a circuit, connect channel 1 to the input signal, and use channel 2 to trace the input signal through the circuit.

Required Reading

Chapters 20 and 21 (Secs. 20-3, 20-5, and 21-2) of *Electronic Principles*, 8th ed.

Equipment

2 power supplies: ±15 V
1 DMM (digital multimeter)
4 diodes: three 1N914, L53RD
4 op amps: three LM741C, LM318
10 ½-W resistors: two 1 kΩ, 2.2 kΩ, 8.2 kΩ, 10 kΩ, two 15 kΩ, 18 kΩ, 22 kΩ, 100 kΩ
5 capacitors: 0.0068 μF, two 0.01 μF, 0.022 μF, 0.047 μF
1 potentiometer: 5 kΩ
1 oscilloscope

Procedure

WIEN-BRIDGE OSCILLATOR

1. Measure and record the resistor and capacitor values. Build the Wien-bridge oscillator.
2. Observe the output of pin 6 with channel 1 of the oscilloscope (5 V/DIV). Adjust the variable 5 kΩ resistor to get a maximum unclipped sinusoidal output.
3. The LED should be lit, indicating an output signal.
4. Measure and record the frequency and the peak-to-peak output voltage:

    $f_{out} =$ _____.

    $v_{out} =$ _____.

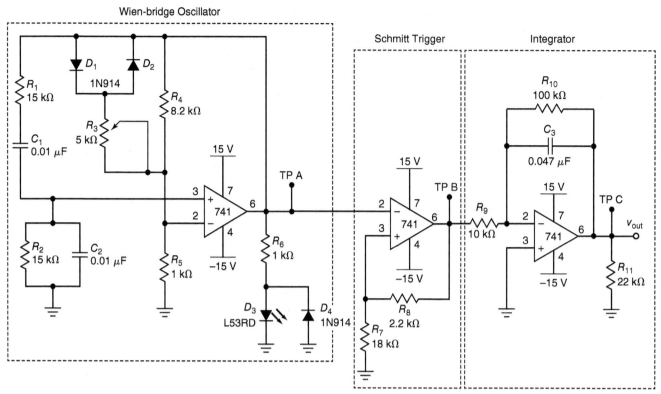

Figure 56-1

5. Sketch the output waveform of the Wien-bridge oscillator, showing the peak-to-peak value and the period:

10. Sketch the input and output waveforms of the Schmitt trigger, showing the peak-to-peak values and the period:

SCHMITT TRIGGER

6. Build the Schmitt trigger. This includes all the circuitry between test points *A* and *B*.
7. Test the Schmitt trigger as follows: Connect the output of the Wien-bridge oscillator to test point *A*. Use channel 1 to monitor the input to the Schmitt trigger (test point *A*), and use channel 2 to monitor the output of the Schmitt trigger (test point *B*).
8. The display on the oscilloscope should show a sine wave and a square wave. Both will have approximately the same peak-to-peak value.
9. Record the frequency and peak-to-peak value of the square wave:

f_{out} = _____.

v_{out} = _____.

INTEGRATOR

11. Build the integrator.
12. Test the integrator as follows: Connect the output of the Schmitt trigger to the input of the integrator.
13. Use channel 1 to monitor the input to the integrator (test point *B*) and use channel 2 to monitor the output of the integrator (test point *C*). A triangular wave should be visible on the oscilloscope.
14. Record the frequency and peak-to-peak value of the triangular wave:

f_{out} = _____.

v_{out} = _____.

15. Sketch the input and output waveforms of the integrator, showing the peak-to-peak values and the period:

16. The triangular wave has a smaller peak-to-peak value than the sine wave and square wave. One way to increase its peak-to-peak value is to use a smaller capacitor.

17. Change the 0.047 μF capacitor to a 0.022 μF capacitor. Record the peak-to-peak value of the triangular wave:

_____.

18. Increase the frequency of the Wien-bridge oscillator by changing the two 0.01 μF capacitors to 0.0047 μF.

19. Notice how the slew rate of a 741C is degrading the shape of the square wave and the triangular wave. Since a 741C has a typical slew rate of 0.5 V/μs, it takes typically 56 μs to slew through a peak-to-peak change of 28 V. Since the period is theoretically 443 μs, the slew-rate distortion is noticeable.

20. Here is one way to improve the performance: Turn the power off and replace the 741C with an LM318.

21. Turn the power back on, and well-defined square waves and triangular waves will be visible on the oscilloscope. Explain why the waveforms look cleaner and sharper:

CRITICAL THINKING

22. Briefly describe how a technician would modify the function generator of Fig. 56-1 to make it tunable over the audio range of 20 Hz to 20 kHz.

23. Suppose the design requires the buffering of the A, B, and C outputs, with a 600 Ω output impedance on each buffered output. Describe one simple way to do this, using three more op amps:

Experiment 56

Lab Partner(s) _____

PARTS USED Nominal Value	Measured Value	PARTS USED Nominal Value	Measured Value	PARTS USED Nominal Value	Measured Value
1 kΩ		15 kΩ		0.0068 μF	
1 kΩ		15 kΩ		0.01 μF	
2.2 kΩ		18 kΩ		0.01 μF	
8.2 kΩ		22 kΩ		0.022 μF	
10 kΩ		100 kΩ		0.047 μF	

Questions for Experiment 56

1. If the supply voltages in Fig. 56-1 are changed to ± 10 V, the peak-to-peak output   ( )
of the Wien-bridge oscillator is closest to:
(a) 10 V;   (b) 20 V;   (c) 30 V;   (d) 40 V.

2. If all diodes are reversed in Fig. 56-1, the: ( )
(a) oscillator will stop;   (b) LED will go out;   (c) LED is destroyed;
(d) circuit works as before.

3. If the 10 kΩ resistance is reduced slightly in Fig. 56-1, the: ( )
(a) square-wave output becomes smaller;   (b) triangular wave becomes
bigger;   (c) LED goes out;   (d) square wave becomes smaller.

4. The Schmitt trigger works better with an LM318 because this op amp has: ( )
(a) less offset voltage;   (b) a higher CMRR;   (c) a higher slew
rate;   (d) an open collector.

5. Reducing the capacitors in the Wien-bridge oscillator will: ( )
(a) decrease the frequency;   (b) increase the frequency;   (c) produce a
faster slew rate;   (d) increase the triangular output.

6. The average current through the LED in Fig. 56-1 is closest to: ( )
(a) 3 mA;   (b) 6 mA;   (c) 12 mA;   (d) 24 mA.

7. If the 2.2 kΩ resistor of Fig. 56-1 is increased, the: ( )
(a) hysteresis is reduced;   (b) peak-to-peak value is increased;
(c) frequency is decreased;   (d) output offset is reduced.

8. To operate the function generator of Fig. 56-1 at higher frequencies: ( )
(a) shorten the wires;   (b) use a better op amp;   (c) use smaller
capacitors;   (d) all of the foregoing.

9. Optional. Instructor's question.

10. Optional. Instructor's question.

6

Power Supply Monitoring Circuit

In this experiment, a window detector with preset limits is used to monitor a +9 Vdc power supply. If the voltage from the power supply falls to 8.82 V (LTP) or rises to 9.28 V (UTP), the output of the window detector will go HIGH. This in turn will cause the transistor in the LED driver to go into saturation and the LED will go out. If the voltage from the power supply remains between +8.82 V and 9.28 V, the output of the window detector will remain LOW. This will cause the transistor in the LED driver to stay in cutoff and the LED will light. The potentiometer R_1 provides the +9 Vdc. A faulty power supply can be simulated by adjusting R_1 to cause the output voltage to vary.

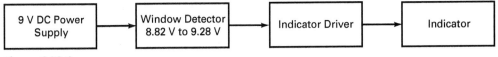

Figure SA6-1

In this experiment, the oscilloscope and DMM will be used to trace dc voltages through a power supply monitoring circuit. This experiment will also provide an opportunity to improve troubleshooting skills.

GOOD TO KNOW

When troubleshooting a circuit, a good technician divides the circuit into logical blocks and then takes voltage measurements to verify that each portion of the circuit is working properly.

Required Reading

Chapters 3, 5, and 20 of *Electronic Principles*, 8th ed.

Equipment

2 power supplies: ±12 V
2 op amps: LM741

1 *npn* transistor: 2N3904
2 diodes: 1N914 or 1N4148
1 L53RD (red LED)
5 ½-W resistors: 1 kΩ, 2.2 kΩ, 2.7 kΩ, two 7.5 kΩ
1 10 kΩ potentiometer
1 DMM (digital multimeter)
1 oscilloscope

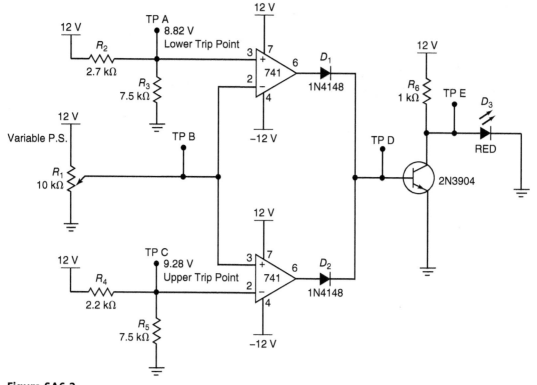

Figure SA6-2

Procedure

WINDOW DETECTOR

1. Using a DMM, measure and record the value of the resistors. Change the DMM mode to Diode Test, and measure and record the voltage across both silicon diodes when forward and reverse biased.

2. Build the power supply monitoring circuit, as shown in Fig. SA6-2.

3. The voltage divider created by potentiometer R_1 will be used to simulate a +9 Vdc power supply. To simulate a properly operating power supply, adjust R_1 until +9 Vdc is present at TP B.

4. Calculate the expected voltages at test points A and C. Record the calculated values in Table SA6-1.

5. Using the DMM, measure the voltages at test points A and C. Record the measured values in Table SA6-1.

6. Since the +9 Vdc present at test point B is within the lower trip point (LTP) and upper trip point (UTP) of the window detector, the voltage at test point D should be a negative voltage. Measure the voltage at test point D and record the value in Table SA6-1.

7. Determine if the transistor is in saturation or cutoff. Record the answer in Table SA6-1.

8. Is the LED ON or OFF? Record the answer in Table SA6-1.

9. While measuring the voltage at test point B, slowly adjust potentiometer R_1 until the voltage drops to 8.82 Vdc (LTP). The LED should go out at this point. If it hasn't, continue decreasing the voltage until it does.

10. Measure the voltage at test point D and record the value in Table SA6-1.

11. Determine if the transistor is in saturation or cutoff. Record the answer in Table SA6-1.

12. Is the LED ON or OFF? Record the answer in Table SA6-1.

13. While measuring the voltage at test point B, slowly adjust potentiometer R_1 until the voltage increases to 9.28 Vdc (UTP). The LED should go out at this point. If it hasn't, continue increasing the voltage until it does.

14. Measure the voltage at test point D and record the value in Table SA6-1.

15. Determine if the transistor is in saturation or cutoff. Record the answer in Table SA6-1.

16. Is the LED ON or OFF? Record the answer in Table SA6-1.

LED DRIVER

17. It is common to have a device or component that requires more current or voltage than the circuit controlling it can provide. The transistor in the LED driver configuration is being switched between saturation and cutoff by the output of the window detector. When the transistor is in cutoff and the LED is lit, the current flowing through it and the voltage dropped across it are provided by the +12 Vdc supply, not the window detector. When the transistor is in saturation, the voltage at the collector is not sufficient to light the LED.

Using the DMM, measure the voltage at test points D and E while adjusting R_1 until the LED is off. Record the measurements in Table SA6-2.

18. Slowly adjust R_1 in the opposite direction until the LED is lit. Using the DMM, measure the voltage at test points D and E. Record the measurements in Table SA6-2.

TROUBLESHOOTING

19. Assume the circuit has the faults listed in Table SA6-3. Record the anticipated circuit behavior for each assumed fault.

20. Insert the faults listed in Table SA6-3 into the circuit and observe how the circuit responds to the varying of R_1. Record how the circuit's operation changed due to the faults.

CRITICAL THINKING

21. R_2 and R_3 form a voltage divider for the LTP. R_4 and R_5 form a voltage divider for the UTP. What resistor values would be needed to create a 0.5 V window centered on 10 Vdc?

System Application 6

Lab Partner(s)_____

PARTS USED Nominal Value	Measured Value	
1 kΩ		
2.2 kΩ		
2.7 kΩ		
7.5 kΩ		
7.5 kΩ		
	V_F	V_R
1N914 or 1N4148		
1N914 or 1N4148		

TABLE SA6-1 DC VOLTAGE MEASUREMENTS AND OBSERVATIONS

TP B = +9 V			
Test Point	Calculated	Multisim	Actual
A			
C			
D			
XSTR Saturation / Cutoff		LED On / Off	

TP B = +8.82 V (LTP)			
Test Point	Calculated	Multisim	Actual
D			
XSTR Saturation / Cutoff		LED On / Off	

TP B = +9.28 V (UTP)			
Test Point	Calculated	Multisim	Actual
D			
XSTR Saturation / Cutoff		LED On / Off	

TABLE SA6-2 LED DRIVER VOLTAGES

Test Point	LED Off	LED On
D		
E		

TABLE SA6-3 TROUBLESHOOTING

Trouble	Observations of how the circuit's operation changed
R_3 shorted	
R_4 shorted	
2N3904 fails	
D_1 opens	
R_6 opens	

Questions for System Application 6

1. R_4 and R_5 form a voltage divider to provide _____ at test point C.
   (a) +4.5 V;   (b) +8.82 V;   (c) +5.2 V;   (d) +9.28 V.
2. R_2 and R_3 form a voltage divider to provide _____ at test point A.
   (a) +4.5 V;   (b) +8.82 V;   (c) +5.2 V;   (d) +9.28 V.
3. The transistor functions as a(n) _____.
   (a) amplifier;   (b) chopper;   (c) switch;   (d) resistance.
4. If R_3 is shorted, the _____ no longer works.
   (a) XSTR;   (b) LTP;   (c) LED;   (d) UTP.
5. If R_4 is shorted, the _____ no longer works.
   (a) XSTR;   (b) LTP;   (c) LED;   (d) UTP.
6. When the XSTR is in cutoff, the LED is _____.
   (a) off;   (b) flickering;   (c) on;   (d) dim.
7. When the XSTR is in saturation, the LED is _____.
   (a) off;   (b) flickering;   (c) on;   (d) dim.
8. When TP B measures +9 V, the XSTR is _____.
   (a) in cutoff;   (b) in saturation;   (c) faulty;   (d) oscillating.
9. When TP B measures +10 V, the XSTR is _____.
   (a) in cutoff;   (b) in saturation;   (c) faulty;   (d) oscillating.
10. When TP B measures +8 V, the XSTR is _____.
    (a) in cutoff;   (b) in saturation;   (c) faulty;   (d) oscillating.

Experiment 57

The 555 Timer

The 555 timer combines a relaxation oscillator, two comparators, and an *RS* flip-flop. This versatile chip can be used as an astable multivibrator, monostable multivibrator, VCO, ramp generator, etc. In this experiment, various 555 timer circuits will be built and analyzed.

GOOD TO KNOW

The 555 timer can be used as a timer that is variable from microseconds to hours. The TTL-compatible output is capable of sinking and sourcing 200 mA.

Required Reading

Chapter 21 (Secs. 21-7 to 21-9) of *Electronic Principles,* 8th ed.

Equipment

1 function generator
1 power supply: 15 V
10 ½-W resistors: two 1 kΩ, 4.7 kΩ, two 10 kΩ, 22 kΩ, 33 kΩ, 47 kΩ, 68 kΩ, 100 kΩ
1 potentiometer: 1 kΩ
4 capacitors: 0.01 μF, 0.1 μF, two 0.47 μF
1 transistor: 2N3906
1 op amp: 741C
1 timer: NE555
1 oscilloscope
1 frequency counter
1 DMM (digital multimeter)

Procedure

ASTABLE 555 TIMER

1. Calculate the frequency and duty cycle in Fig. 57-1 for the resistances listed in Table 57-1. Record your answers.
2. Measure and record the resistor and capacitor values. Build the circuit shown in Fig. 57-1 with $R_1 = 10$ kΩ and $R_2 = 100$ kΩ.

3. Observe the output with an oscilloscope. Measure and record the frequency.
4. Measure W. Calculate and record the duty cycle as the measured D in Table 57-1.
5. Observe the voltage across the timing capacitor (pin 6). An exponentially rising and falling wave between 5 and 10 V should be visible on the oscilloscope.
6. Repeat Steps 2 through 5 for the other resistances of Table 57-1.

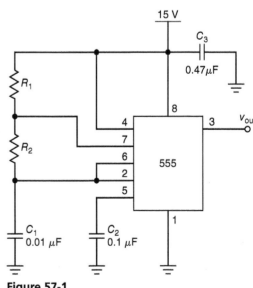

Figure 57-1

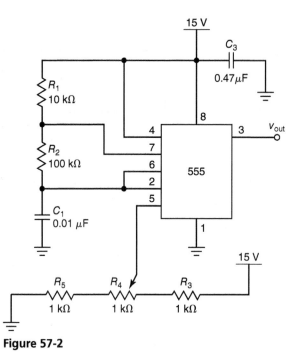

Figure 57-2

VOLTAGE-CONTROLLED OSCILLATOR

7. Build the VCO shown in Fig. 57-2.
8. Observe the output with an oscilloscope. Vary the potentiometer and notice what happens. Record the minimum and maximum frequencies in Table 57-2.
9. Disconnect pin 5 from the potentiometer and couple a 10 Hz signal into this pin. Slowly increase the level of this signal from zero. Notice the horizontal jitter that appears on the signal. This is the PPM discussed in Sec. 21-9 of the textbook.

MONOSTABLE 555 TIMER

10. Figure 57-3 shows a Schmitt trigger driving a monostable 555 timer. Assume that it produces a normal trigger input for the 555. Calculate and record the pulse width out of the 555 timer for each R listed in Table 57-3.
11. Build the circuit shown in Fig. 57-3 with an R of 33 kΩ.
12. Observe the output of the Schmitt trigger (pin 6 of the 741C). Set the frequency of the sine-wave input to 1 kHz. Adjust the sine-wave level until the Schmitt-trigger output has a duty cycle of approximately 90 percent.

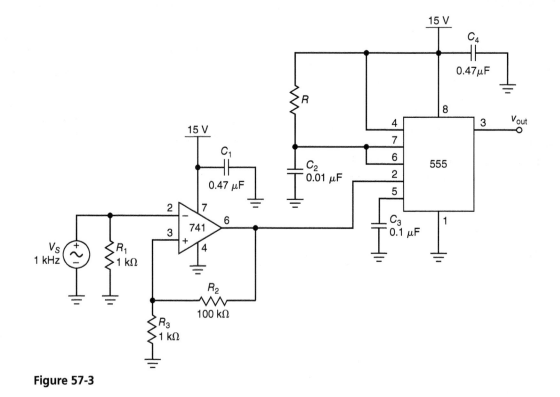

Figure 57-3

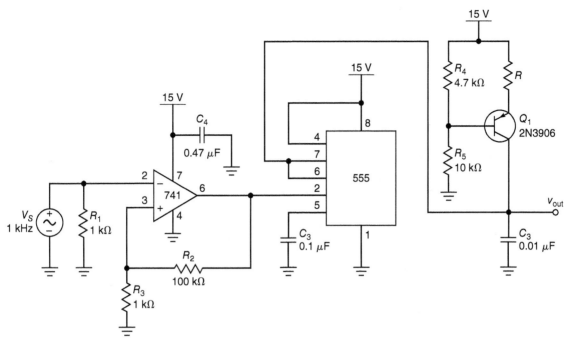

Figure 57-4

13.
Observe the output of the 555 timer. Measure and record the pulse width.

14.
Repeat Steps 11 through 13 for the remaining R values of Table 57-3.

RAMP GENERATOR

15.
Figure 57-4 shows a ramp generator. As before, the Schmitt trigger drives a 555 timer connected for monostable operation. But now the timing capacitor is charged by a *pnp* current source rather than a resistor. For each value of R listed in Table 57-4, calculate the slope of the output waveform.

16.
Build the circuit shown in Fig. 57-4 with an R of 10 kΩ.

17.
Set the ac generator to 1 kHz. Adjust the level to get a duty cycle of approximately 90 percent out of the Schmitt trigger.

18.
Observe the output voltage; it should be a positive ramp. Measure the ramp voltage and time. Then work out the slope. Record the value in Table 57-4.

19.
Repeat Steps 16 through 18 for the remaining values of R (Table 57-4).

TROUBLESHOOTING

20.
Assume that R equals 22 kΩ in Fig. 57-4. Here are the symptoms: (1) no ramp appears at the final output; (2) a normal Schmitt-trigger output drives pin 2 of the 555 timer; (3) approximately +10 V appears at the base of the 2N3906. Try to figure out what troubles (there is more than one possibility) can cause these symptoms. Insert each suspected trouble to verify that it does cause the symptoms. Record all the troubles located (Table 57-5).

CRITICAL THINKING

21.
Select a value of C_3 that produces a slope of 15 V/ms when R equals 10 kΩ.

22.
Build the circuit with an R of 10 kΩ and the design value of C_3. Measure the slope of the output signal. Record the quantities of Table 57-6.

333

Experiment 57

Lab Partner(s) _____

PARTS USED Nominal Value	Measured Value	PARTS USED Nominal Value	Measured Value	PARTS USED Nominal Value	Measured Value
1 kΩ		22 kΩ		0.01 μF	
1 kΩ		33 kΩ		0.1 μF	
4.7 kΩ		47 kΩ		0.47 μF	
10 kΩ		68 kΩ		0.47 μF	
10 kΩ		100 kΩ			

TABLE 57-1. ASTABLE MULTIVIBRATOR

		Calculated		Measured			
				Multisim		Actual	
R_1	R_2	f	D	f	D	f	D
10 kΩ	100 kΩ						
100 kΩ	10 kΩ						
10 kΩ	10 kΩ						

TABLE 57-2. VCO OPERATION

	Multisim	Actual
f_{min}		
f_{max}		

TABLE 57-3. MONOSTABLE MULTIVIBRATOR

		Measured W	
R	Calculated W	Multisim	Actual
33 kΩ			
47 kΩ			
68 kΩ			

TABLE 57-4. RAMP GENERATOR

		Measured slope	
R	Calculated slope	Multisim	Actual
10 kΩ			
22 kΩ			
33 kΩ			

TABLE 57-5. TROUBLESHOOTING

	Description	
Trouble	Multisim	Actual
1		
2		
3		

TABLE 57-6. CRITICAL THINKING: C_3 = _____

	Multisim	% Error	Actual	% Error
Slope				

Questions for Experiment 57

1. In Fig. 57-1, the calculated frequency for R_1 and R_2 (both equal to 10 kΩ) is approximately: ( )
   (a) 686 Hz;   (b) 1.2 kHz;   (c) 4.8 kHz;   (d) 6.91 kHz.
2. In Fig. 57-2, the adjustment controls the: ( )
   (a) output frequency;   (b) output voltage;   (c) supply voltage;
   (d) input voltage.
3. The output of the Schmitt trigger in Fig. 57-3 is: ( )
   (a) always positive;   (b) always negative;   (c) positive on one half-cycle
   and negative on the other;   (d) a constant dc voltage.
4. An R of 47 kΩ in Fig. 57-3 produces a pulse that is closest to: ( )
   (a) 363 μs;   (b) 517 μs;   (c) 748 μs;   (d) 1000 μs.
5. In Fig. 57-4, an R of 10 kΩ produces a slope of approximately: ( )
   (a) 12.4 V/ms;   (b) 18.6 V/ms;   (c) 41 V/ms;   (d) 56 V/ms.
6. In Fig. 57-3, how much input voltage is needed to get an output from the Schmitt trigger? Why?

7. Briefly describe the circuit operation of Fig. 57-4.

TROUBLESHOOTING

8. In Fig. 57-4, assume that R equals 10 kΩ and the output slope is 410 V/ms. Name a trouble that can produce this slope.

CRITICAL THINKING

9. How were the design values determined?

10. Optional. Instructor's question.

555 Timer Applications

This experiment will provide additional insight into the 555, a widely used timing circuit. The first application is a START and RESET circuit. Pushing a START button triggers a mono-stable circuit. The circuit can be reset at any time by pushing the RESET button. The second application is an ALARM circuit. Pushing the ALARM switch turns on an astable multivibrator that drives a speaker. The third application is a pulse-position modulator that uses two 555 timers. The first timer modulates the second, the output of which drives the speaker.

GOOD TO KNOW

Many circuits have been designed around the 555 timer. A search on the Internet will produce a wide variety of circuits that utilize the 555 timer.

Required Reading

Chapter 21 (Secs. 21-7 and 21-9) of *Electronic Principles,* 8th ed.

Equipment

- 1 power supply: 12 V
- 1 DMM (digital multimeter)
- 2 diodes: L53RD, LR53GD
- 2 timers: two LM555
- 11 ½-W resistors: 100 Ω, 470 Ω, 1 kΩ, 1.2 kΩ, 1.8 kΩ, three 10 kΩ, 100 kΩ, two 1 MΩ
- 8 capacitors: two 0.01 µF, 0.1 µF, 0.33 µF, 1 µF, two 10 µF, 470 µF
- 1 8 Ω speaker (small)
- 1 oscilloscope

Procedure

START AND RESET

1. Figure 58-1 is the START-and-RESET circuit discussed in Sec. 21-9 of the textbook. The START button controls the beginning of a timed interval of high output voltage. The RESET button can be used to terminate the interval if desired.

2. Measure and record the resistor and capacitor values. Build the circuit shown in Fig. 58-1. Set the 10 kΩ variable resistance to minimum. Use push-button switches if available. If not, use SPST switches or temporary jumper connections when called for.

3. Push the START switch and release it. (If SPST switches are being used, close the switch temporarily and then open it. If jumper wires are being used, temporarily connect a wire between pin 2 and ground and then disconnect it.)

4. The LED will light up. It will stay on for an interval of time. Then it will go out.

5. Use channel 1 to observe the charging waveform on pin 6, and use channel 2 for the output on pin 3.

6. Repeat Step 3 and watch the waveforms on the oscilloscope. With a stopwatch or by counting 1001, 1002, 1003, and so on, determine how long the LED stays on. Record the time in seconds:

   _____.

7. Set the 10 kΩ variable resistor to maximum. Repeat Step 6 and record the time:

   _____.

8. Steps 8 through 13 are optional. Connect a buzzer (MCM Part#: 28-740) between pin 3 and ground. The buzzer will be in parallel with the LED branch.

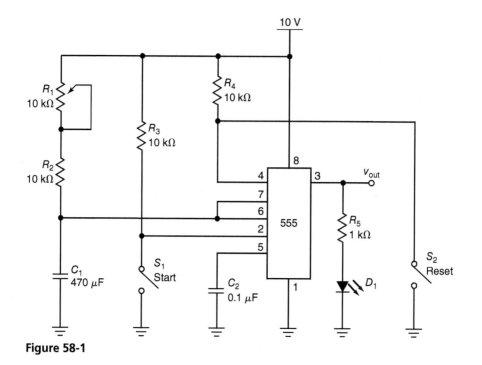

Figure 58-1

9. Repeat Step 3. The buzzer will sound, and the LED will light for a timed interval. Then, both will become inactive.

10. Disconnect the buzzer and the LED branch so that nothing is connected to pin 3.

11. Add the emitter follower and motor (MCM Part#: 28-12810) of Fig. 58-2 to the circuit.

12. Repeat Step 3. The motor will run for a timed interval and then stop.

13. What was learned in Steps 3 through 12?

18. Steps 18 through 20 are optional. Connect a photocell (Digikey Part #: PDV-P9008-ND) in parallel with the ALARM switch. Point the photocell toward a light source.

19. Open the ALARM switch. Cover the photocell to darken it. The speaker will produce a sound. (*Note:* If the speaker does not produce a sound, the photocell resistance is too high. Try again after changing the 10 kΩ resistor on pin 4 to 100 kΩ.)

20. What was learned in Steps 14 through 19?

SIRENS AND ALARMS

14. Fig. 58-3 shows a circuit that can sound an alarm when the ALARM switch is opened. It is an astable multivibrator, waiting for someone or something to open the alarm switch.

15. Build the circuit.

16. When power is applied, open the ALARM switch, and a sound will be heard from the speaker.

17. Vary the 1 MΩ resistor to change the frequency.

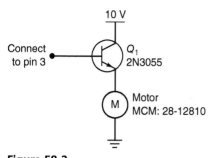

Figure 58-2

VCO WITH PULSE-POSITION MODULATION

21. In Fig. 58-4, both stages are astable multivibrators. Calculate the oscillation frequency for each stage:

$f_1 = $ _____ . $f_2 = $ _____ .

22. With the switch in position 2, only the frequency of the second stage reaches the speaker. But when the switch is moved to position 1, the output of the first stage is used to modulate the second stage. This will produce a warbled output from the second stage, a sound that rises and falls in frequency.

23. Build all of the circuits shown in Fig. 58-4, except for the speaker.

24. After power is applied, the green LED should be flashing slowly. This indicates that stage 1 is oscillating at a very low frequency.

25. Make sure that the switch in the second stage is in position 2. Look at the output of stage 2 with an oscilloscope to make sure that it is oscillating at approximately 900 Hz. After synchronizing the signal, a rectangular wave with a duty cycle of about 60 percent should be visible on the oscilloscope.

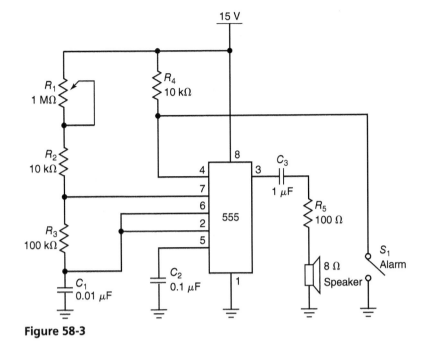

Figure 58-3

26. Continue to monitor the output of the second stage with the oscilloscope. Change the switch in the second stage to position 1, so that the output of the first stage can modulate the output of the second stage.

27. Describe what was observed with regard to the modulated output frequency, as seen on the oscilloscope:

28. Move the switch back to position 2 to remove the modulation. Connect a small 8 Ω speaker to the output, as shown in Fig. 58-4. The speaker will produce a tone with a frequency of approximately 900 Hz.

29. Move the switch to position 1 to get modulation. Now, a warbled output sound will be audible.

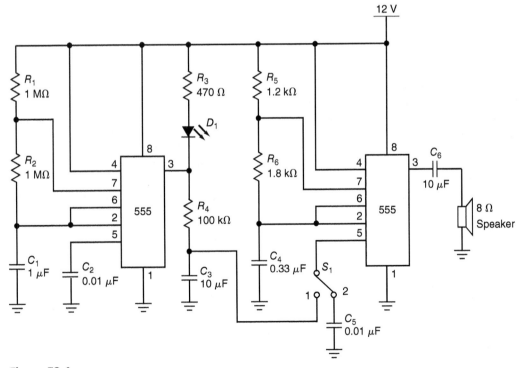

Figure 58-4

Experiment 58

Lab Partner(s) _____

PARTS USED Nominal Value	Measured Value	PARTS USED Nominal Value	Measured Value	PARTS USED Nominal Value	Measured Value
100 Ω		10 kΩ		0.33 μF	
470 Ω		100 kΩ		1 μF	
1 kΩ		1 MΩ		10 μF	
1.2 kΩ		1 MΩ		10 μF	
1.8 kΩ		0.01 μF		470 μF	
10 kΩ		0.01 μF			
10 kΩ		0.1 μF			

Questions for Experiment 58

1. The circuit of Fig. 58-1 is: ( )
   (a) monostable;   (b) astable;   (c) bistable;   (d) unstable.
2. The circuit of Fig. 58-1 has a maximum pulse width of approximately: ( )
   (a) 1 s;   (b) 2 s;   (c) 5 s;   (d) 10 s.
3. If the motor of Fig. 58-2 has a current of 1 A, the base current is closest to: ( )
   (a) 1 μA;   (b) 2 mA;   (c) 200 mA;   (d) 1 A.
4. The circuit of Fig. 58-3 is: ( )
   (a) monostable;   (b) astable;   (c) bistable;   (d) ultrastable.
5. The circuit of Fig. 58-3 has a maximum frequency of approximately: ( )
   (a) 123 Hz;   (b) 686 Hz;   (c) 5 kHz;   (d) 11.2 kHz.
6. The first stage of Fig. 58-4 has an output frequency of approximately: ( )
   (a) 0.46 Hz;   (b) 2.3 Hz;   (c) 123 Hz;   (d) 909 Hz.
7. The second stage of Fig. 58-4 has an output frequency of approximately: ( )
   (a) 0.48 Hz;   (b) 2.3 Hz;   (c) 454 Hz;   (d) 909 Hz.
8. When the LED is turned on in Fig. 58-4, the LED current is approximately: ( )
   (a) 10 mA;   (b) 20 mA;   (c) 40 mA;   (d) 80 mA.
9. If both timing resistances are doubled in the second stage of Fig. 58-4, the output   ( )
   frequency is closest to:
   (a) 0.48 Hz;   (b) 2.3 Hz;   (c) 454 Hz;   (d) 909 Hz.
10. Optional. Instructor's question.

Shunt Regulators

The shunt regulator works fine as long as the changes in line and load are not too great. It has the advantage of a simple circuit design that includes built-in short-circuit protection. It has the disadvantage of poor efficiency because of the power wasted in the series resistor. In this experiment, three different shunt regulators will be built and tested.

GOOD TO KNOW

Voltage regulators are evaluated on their ability to maintain a constant output voltage while the input voltage varies (line regulation) and the load resistance varies (load regulation).

Required Reading

Chapter 22 (Secs. 22-1 and 22-2) of *Electronic Principles,* 8th ed.

Equipment

2 power supplies: ± 15 V
1 DMM (digital multimeter)
2 diodes: 1N5234B, 1N757
2 transistors: 2N3904, 2N3055
1 op amp: 741C
13 ½-W resistors: two 100 Ω, 220 Ω, three 330 Ω, 470 Ω, two 1 kΩ, 2.2 kΩ, 3.3 kΩ, 5.6 kΩ, 6.8 kΩ
2 potentiometers: 1 kΩ, 5 kΩ
1 oscilloscope

Procedure

FIXED SHUNT REGULATOR

1. In Fig. 59-1, half-watt resistors are used in parallel to avoid the need for 1-W resistors. The parallel 100 Ω resistors produce an equivalent series resistance of 50 Ω, and the parallel 330 Ω resistors produce an equivalent load resistance of 110 Ω.
2. Measure and record the resistor values. Build the circuit shown in Fig. 59-1.
3. Use a DMM to measure the output voltage across the 330 Ω resistors. Record this value as the full-load voltage:

$V_{FL} =$ _____.

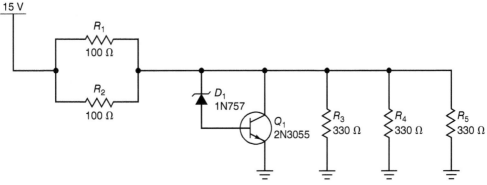

Figure 59-1

4. Remove one of the 330 Ω resistors and record the output voltage:

_____.

5. Remove another 330 Ω resistor and record the output voltage:

_____.

6. Remove the last 330 Ω resistor and record this as the no-load voltage:

V_{NL} = _____.

7. Calculate and record the load regulation:

_____.

8. Reconnect the three 330 Ω resistors.
9. Increase the input voltage to +16.5 V. Record the output voltage as the high-line output:

V_{HL} = _____.

10. Reduce the input voltage to +13.5 V. Record the output voltage as the low-line output:

V_{LL} = _____.

11. Watch the DMM reading of output voltage while the input voltage is reduced. Notice how the regulator stops regulating when there is insufficient input voltage.

12. Calculate and record the line regulation:

_____.

ADJUSTABLE SHUNT REGULATOR

13. Build the circuit shown in Fig. 59-2.
14. Use a DMM to measure and record the zener voltage:

V_Z = _____.

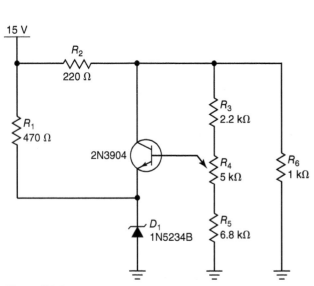

Figure 59-2

15. Measure the load voltage across the 1 kΩ resistor. Vary the 5 kΩ potentiometer in both directions to record the minimum and maximum load voltages:

$V_{L(max)}$ = _____.

$V_{L(min)}$ = _____.

16. Adjust the potentiometer to get a load voltage of 10 V. Record this value as the full-load voltage:

V_{FL} = _____.

17. Remove the 1 kΩ load resistor and record this as the no-load voltage:

V_{NL} = _____.

18. Calculate and record the load regulation:

_____.

19. Reconnect the 1 kΩ load resistor.
20. Increase the input voltage to +16.5 V. Record the output voltage as the high-line output:

V_{HL} = _____.

21. Reduce the input voltage to +13.5 V. Record the output voltage as the low-line output:

V_{LL} = _____.

22. Watch the DMM reading of output voltage while the input voltage is reduced. Notice how the regulator stops regulating when there is insufficient input voltage.

23. Calculate and record the line regulation:

_____.

OP-AMP SHUNT REGULATOR

24. Build the circuit shown in Fig. 59-3.
25. Use a DMM to measure and record the zener voltage:

V_Z = _____.

26. Measure the load voltage across the 1 kΩ resistor. Vary the 1 kΩ potentiometer in both directions to record the minimum and maximum load voltages:

$V_{L(max)}$ = _____.

$V_{L(min)}$ = _____.

27. Adjust the potentiometer to get a load voltage as close as possible to 10 V. Record this value as the full-load voltage:

V_{FL} = _____.

28. Remove the 1 kΩ load resistor and record this as the no-load voltage:

V_{NL} = _____.

29. Calculate and record the load regulation:

_____.

30. Reconnect the 1 kΩ load resistor.
31. Increase the input voltage to +16.5 V. Record the output voltage as the high-line output:

V_{HL} = _____.

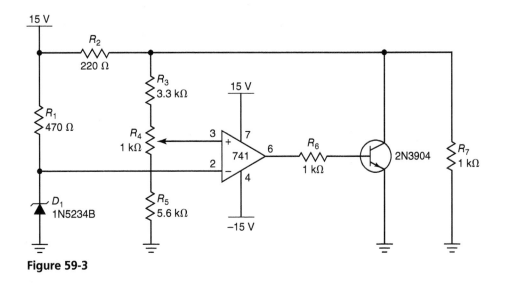

Figure 59-3

32. Reduce the input voltage to +13.5 V. Record the output voltage as the low-line output:

$V_{LL} =$ _____.

33. Calculate and record the line regulation: _____.

ADDITIONAL WORK (OPTIONAL)

34. Graph the load voltage versus the load resistance for the circuit of Fig. 59-1 using Steps 3 through 6.

35. Graph the load voltage versus the load current for the circuit of Fig. 59-1 using Steps 3 through 6.

36. Have another student insert the following trouble into any of the three circuits: open any component or connecting wire. Use only voltage readings of a DMM to troubleshoot.

37. Repeat Step 36 several times to build confidence in being able to troubleshoot the circuit for various troubles.

Experiment 59

Lab Partner(s) _____

PARTS USED Nominal Value	Measured Value	PARTS USED Nominal Value	Measured Value	PARTS USED Nominal Value	Measured Value
100 Ω		330 Ω		3.3 kΩ	
100 Ω		470 Ω		5.6 kΩ	
220 Ω		1 kΩ		6.8 kΩ	
330 Ω		1 kΩ		1 kΩ Pot	
330 Ω		2.2 kΩ		5 kΩ Pot	

Questions for Experiment 59

1. When the load current increases in Fig. 59-1, the zener current: (   )
   (a) decreases;   (b) increases;   (c) stays the same.
2. In Fig. 59-1, the input current to the shunt regulator is approximately: (   )
   (a) 25 mA;   (b) 50 mA;   (c) 100 mA;   (d) 200 mA.
3. If the 2N3055 has a current gain of 50 in Fig. 59-1, the zener current with no   (   )
   load is:
   (a) 0;   (b) 1 mA;   (c) 2 mA;   (d) 5 mA.
4. When the wiper is moved down in Fig. 59-2, the load voltage: (   )
   (a) decreases;   (b) increases;   (c) stays the same.
5. In Fig. 59-2, the zener current is closest to: (   )
   (a) 20 mA;   (b) 30 mA;   (c) 40 mA;   (d) 50 mA.
6. When the wiper is moved up in Fig. 59-3, the load voltage: (   )
   (a) decreases;   (b) increases;   (c) stays the same.
7. One way to increase the maximum load current in Fig. 59-3 is to use a 2N3055   (   )
   and to decrease the:
   (a) 220 Ω;   (b) 470 Ω;   (c) 2.2 kΩ;   (d) 6.8 kΩ.
8. When load current increases in Fig. 59-3, the zener current: (   )
   (a) decreases;   (b) increases;   (c) stays the same.
9. If the load resistance decreases in Fig. 59-3, the output voltage of the op amp: (   )
   (a) decreases;   (b) increases.
10. Optional. Instructor's question.

Series Regulators

The dc voltage out of a bridge rectifier has a peak-to-peak ripple typically around 10 percent of the unregulated dc voltage. By using this unregulated voltage as the input to a voltage regulator, the dc output voltage can be held almost constant with very little ripple. A voltage regulator uses noninverting voltage feedback. The input or reference voltage comes from a zener diode. This zener voltage is amplified by the closed-loop voltage gain of the regulator. The result is a larger dc output voltage with the same temperature coefficient as the zener diode. Most voltage regulators include current limiting to prevent an accidental short across the load terminals from destroying the pass transistor or diodes in the unregulated supply.

GOOD TO KNOW

Transistors are used in voltage regulator circuits because of their ability to conduct large amounts of current. The pinouts for the transistors can be found on the Internet by searching on their model numbers.

Required Reading

Chapter 22 (Secs. 22-1 to 22-3) of *Electronic Principles,* 8th ed.

Equipment

1 power supply: adjustable from 0 to 15 V
11 ½-W resistors: 100 Ω, 220 Ω, 330 Ω, 470 Ω, two 680 Ω, 1 kΩ, two 2.2 kΩ, 4.7 kΩ, 10 kΩ
1 zener diode: 1N5234B
3 transistors: 2N3904
1 capacitor: 0.1 μF
1 DMM (digital multimeter)

Procedure

MINIMUM AND MAXIMUM LOAD VOLTAGE

1. In Fig. 60-1, what is the approximate voltage across the zener diode? Record in Table 60-1. (*Note:* A bypass capacitor of 0.1 μF is used to prevent oscillations.)

2. When R_5 is varied, the load voltage changes in Fig. 60-1. Calculate and record the minimum and maximum load voltage.
3. Measure and record the resistor and capacitor values. Build the circuit shown in Fig. 60-1.
4. Adjust the dc input voltage V_S to $+15$ V. Measure and record the zener voltage.
5. Adjust R_5 to get the minimum load voltage. Measure and record $V_{L(min)}$.
6. Adjust R_5 to get the maximum load voltage. Measure and record $V_{L(max)}$.

LOAD REGULATION

7. Adjust R_5 to get a load voltage of 10 V. Record this as the no-load voltage in Table 60-2.
8. Connect a load resistance of 1 kΩ. Measure the load voltage. Record this as the full-load voltage in Table 60-2.
9. Calculate and record the percent load regulation.

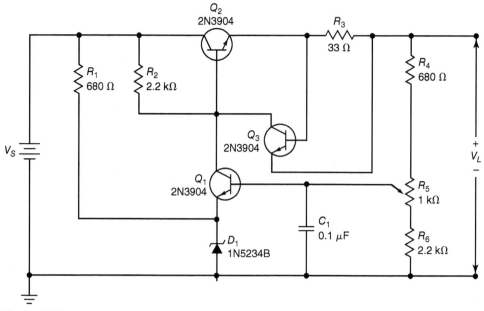

Figure 60-1

LINE REGULATION

10. Measure the load voltage. Record this as $V_{L(\text{max})}$ under "Source regulation" in Table 60-2.
11. Decrease the input voltage from +15 to +12 V. This represents a line change of approximately 20 percent. Measure and record the load voltage as $V_{L(\text{min})}$.
12. Return the input voltage to +15 V. Calculate and record the percent source regulation.

CURRENT LIMITING

13. Assume that the load voltage is 10 V and that Q_3 turns on when V_{BE} is 0.7 V. Notice that the R_4-R_5-R_6 voltage divider has some current through it. Calculate the load current where current limiting begins in Fig. 60-1. Record this at the top of Table 60-3.
14. Connect an R_L of 10 kΩ. Adjust the load voltage to 10 V. Then measure and record the load voltage for each load resistance listed in Table 60-3.
15. Connect a load resistance of 1 kΩ. Short the load terminals and notice how the load voltage goes to zero. Remove the short from the load terminals and notice how the load voltage returns to normal.
16. Use the DMM as an ammeter. Connect the DMM directly across the load terminals. It now measures the load current with a shorted load. This reading should be in the vicinity of the calculated I_{SL} at the top of Table 60-3.

TROUBLESHOOTING

17. For each trouble listed in Table 60-4, estimate and record the load voltage.
18. Insert each trouble into the circuit. Measure and record the load voltage.

CRITICAL THINKING

19. Select a value for R_4 to get a theoretical load voltage of approximately 9 to 13.2 V.
20. Build the circuit with the selected value of R_4.
21. Measure the minimum and maximum load voltage. Record all quantities listed in Table 60-5.

ADDITIONAL WORK (OPTIONAL)

22. Build the circuit shown in Fig. 60-2.
23. Use a DMM to measure and record the zener voltage:

   $V_Z = $ _____.

24. Measure the load voltage across the 1 kΩ resistor. Vary the 1 kΩ potentiometer in both directions to record the minimum and maximum load voltages:

   $V_{L(\text{max})} = $ _____.

   $V_{L(\text{min})} = $ _____.

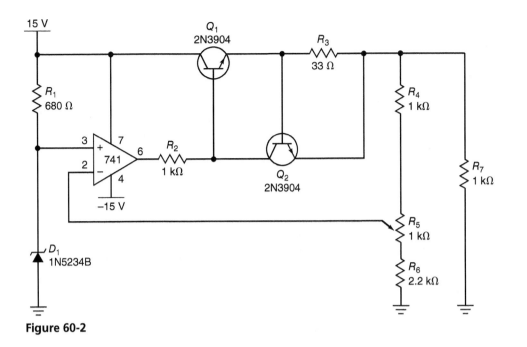

Figure 60-2

25. Adjust the potentiometer to get a load voltage of 10 V. Record this value as the full-load voltage:

    $V_{FL} =$ _____.

26. Remove the 1 kΩ load resistor and record this as the no-load voltage:

    $V_{NL} =$ _____.

27. Calculate and record the load regulation:

    _____.

28. Reconnect the 1 kΩ load resistor.

29. Increase the input voltage to +16.5 V. Record the output voltage as the high-line output:

    $V_{HL} =$ _____.

30. Reduce the input voltage to +13.5 V. Record the output voltage as the low-line output:

    $V_{LL} =$ _____.

31. Calculate and record the load regulation:

    _____.

32. Watch the DMM reading of output voltage while the input voltage is reduced. Notice how the regulator stops regulating when there is insufficient input voltage.

33. Adjust the input voltage to 15 V. Connect an ammeter across the load to produce the shorted-load current. Record the value:

    $I_{SL} =$ _____.

34. What is the power dissipation in the pass transistor when a short is across the load?

    $P_D =$ _____.

35. Briefly describe how to modify the circuit shown in Fig. 60-2 to supply a regulated 10 V across a 10 Ω load resistor:

Experiment 60

Lab Partner(s) _____

PARTS USED Nominal Value	Measured Value	PARTS USED Nominal Value	Measured Value	PARTS USED Nominal Value	Measured Value
100 Ω		680 Ω		2.2 kΩ	
220 Ω		680 Ω		4.7 kΩ	
330 Ω		1 kΩ		10 kΩ	
470 Ω		2.2 kΩ		0.1 μF	

TABLE 60-1. MINIMUM AND MAXIMUM LOAD VOLTAGE

	Calculated	Measured	
		Multisim	Actual
V_Z			
$V_{L(min)}$			
$V_{L(max)}$			

TABLE 60-2. REGULATION

	Load Regulation			Source Regulation		
	V_{NL}	V_{FL}	% LR	$V_{L(max)}$	$V_{L(min)}$	% SR
Multisim						
Actual						

TABLE 60-3. CURRENT LIMITING

R_L	V_L Multisim	V_L Actual
10 kΩ		
4.7 kΩ		
1 kΩ		
470 Ω		
330 Ω		
220 Ω		
100 Ω		
0		

TABLE 60-4. TROUBLESHOOTING

Trouble	Estimated V_{out}	Measured V_{out}	
		Multisim	Actual
R_2 open			
Zener open			
Zener shorted			
Q_1 open			

TABLE 60-5. CRITICAL THINKING: $R_4 =$ _____

	Multisim	% Error	Actual	% Error
$V_{L(min)}$				
$V_{L(max)}$				

Questions for Experiment 60

1. The zener voltage of Fig. 60-1 is approximately: ( )
   (a) 5 V;   (b) 6.2 V;   (c) 7.5 V;   (d) 15 V.
2. Theoretically, the maximum regulated load voltage of Fig. 60-1 is approximately: ( )
   (a) 6.2 V;   (b) 8.37 V;   (c) 12.2 V;   (d) 15 V.
3. Current limiting of Fig. 60-1 starts near: ( )
   (a) 1 mA;   (b) 2.25 mA;   (c) 12.5 mA;   (d) 18.6 mA.
4. The data of Table 60-2 show that load voltage is: ( )
   (a) dependent on load current;   (b) proportional to source voltage;
   (c) almost constant;   (d) low.
5. When V_L is +10 V and R_L is 1 kΩ in Fig. 60-1, the power dissipation in the pass   ( )
   transistor is approximately:
   (a) 50 mW;   (b) 100 mW;   (c) 200 mW;   (d) 279 mW.
6. Briefly explain how the voltage regulator of Fig. 60-1 works.

7. Assume the load terminals are shorted in Fig. 60-1. If the source voltage is +15 V, what is the power dissipation in the pass transistor?

TROUBLESHOOTING

8. Why does the load voltage approach the source voltage when the zener diode opens in Fig. 60-1?

CRITICAL THINKING

9. Suppose the current limiting is required to start at approximately 100 mA in Fig. 60-1. What changes are necessary?

10. Optional. Instructor's question.

Three-Terminal IC Regulators

The LM78XX series is typical of the IC voltage regulators currently available. These three-terminal regulators are the ultimate in simplicity. These voltage regulators are virtually indestructible because of the thermal shutdown discussed in the textbook. In this experiment, an LM7805 will be used as a voltage regulator and a current regulator.

GOOD TO KNOW

When ac coupled, the oscilloscope will only display the ac waveform and will ignore any dc level the ac waveform is riding on. AC coupling is often used to view a small ac ripple riding on a large dc voltage.

Required Reading

Chapter 22 (Sec. 22-4) of *Electronic Principles,* 8th ed.

Equipment

1 signal generator
1 power supply: adjustable from 0 to 15 V
1 voltage regulator: LM7805
6 ½-W resistors: 10 Ω, 22 Ω, 33 Ω, 47 Ω, 68 Ω, 150 Ω
2 capacitors: 0.1 μF, 0.22 μF
1 DMM (digital multimeter)
1 oscilloscope

Procedure

VOLTAGE REGULATOR

1. In Fig. 61-1, estimate and record the output voltage for each input voltage listed in Table 61-1.
2. Measure and record the resistor and capacitor values. Build the circuit shown in Fig. 61-1.
3. Measure and record the output voltage for each input voltage listed in Table 61-1.

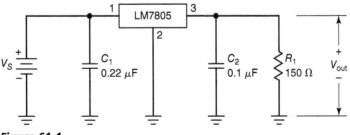

Figure 61-1

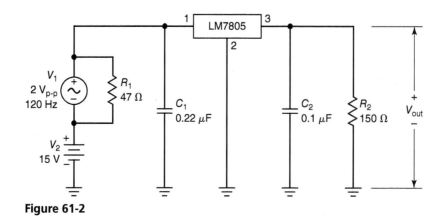

Figure 61-2

RIPPLE REJECTION

4. In Fig. 61-2, the ac source in series with the dc source simulates ripple superimposed on dc voltage. The data sheet of an LM7805 lists the following ripple rejection: minimum is 62 dB and typical is 80 dB. Calculate the peak-to-peak ac output for the minimum and typical ripple rejection. Record the answers in Table 61-2.

5. Build the circuit shown in Fig. 61-2.

6. Observe the ac voltage at the input to the regulator. Adjust the signal source to get 2 V_{p-p} at 120 Hz.

7. Set the oscilloscope to ac coupling. Use the most sensitive ac ranges of the oscilloscope to look at the output ripple. Measure and record this ac output voltage. Then calculate and record the ripple rejection ratio in decibels.

ADJUSTABLE VOLTAGE REGULATOR AND CURRENT REGULATOR

8. The circuit of Fig. 61-3 can function either as a voltage regulator if the output voltage is used or as a current regulator if R_2 is the load resistor. Calculate and record V_{out} and I_{out} for each value of R_2 listed in Table 61-3.

9. Build the circuit with an R_2 of 10 Ω. Measure and record V_{out} and I_{out}.

10. Repeat Step 9 for the other values of R_2.

TROUBLESHOOTING

11. Assume that the circuit of Fig. 61-3 has an R_2 of 68 Ω. For each trouble in Table 61-4, estimate and record the dc output voltage.

12. Build the circuit with an R_2 of 68 Ω. Insert each trouble. Measure and record the dc output voltage.

CRITICAL THINKING

13. Select a value of R_2 in Fig. 61-3 to produce an output voltage of approximately 9 V.

14. Insert the selected value of R_2. Measure the output voltage. Record R_2 and V_{out} in Table 61-5.

APPLICATION (OPTIONAL)—REGULATED POWER SUPPLY

15. In Fig. 61-4, the secondary voltage is 12.6 VAC. Calculate the ideal dc input voltage and ripple to the LM7805:

$V_{dc} =$ _____.

$V_{rip} =$ _____.

16. Build the circuit shown in Fig. 61-4 with the same power transformer used in Experiment 8. Also, use 1N4001s.

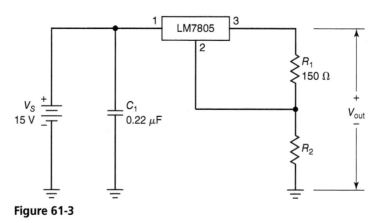

Figure 61-3

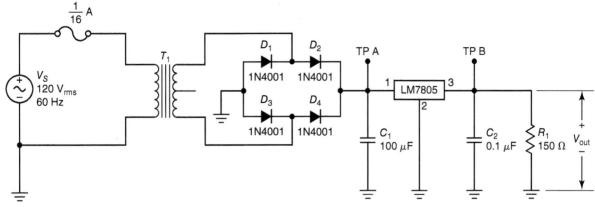

Figure 61-4

17. Measure and record the dc input voltage and ripple to the LM7805:

$V_{dc} = $ _____.

$V_{rip} = $ _____.

18. Measure and record the dc output voltage and ripple of the LM7805:

$V_{dc} = $ _____.

$V_{rip} = $ _____.

19. Compare the input and output values of Steps 17 and 18 and then discuss why these values occur:

DC-TO-DC CONVERTER (OPTIONAL)

20. Section 22-6 of the textbook discusses dc-to-dc converters. In this demonstration, the dc-to-dc converter of Fig. 61-5 will be built and its operation observed.

An astable 555 multivibrator drives the 8 Ω side of the audio transformer. The 600 Ω side drives a half-wave rectifier with a capacitor-input filter.

21. Build the dc-to-dc converter shown in Fig. 61-5.
22. Observe the waveform at pin 3. Record the peak-to-peak value:

_____.

23. Measure the dc voltage across the final 100 kΩ resistor with a DMM. Record the value:

_____.

24. Use the oscilloscope on ac coupling to measure the peak-to-peak ripple across the load resistor:

_____.

25. Explain why the final output voltage is a dc voltage larger than the supply voltage:

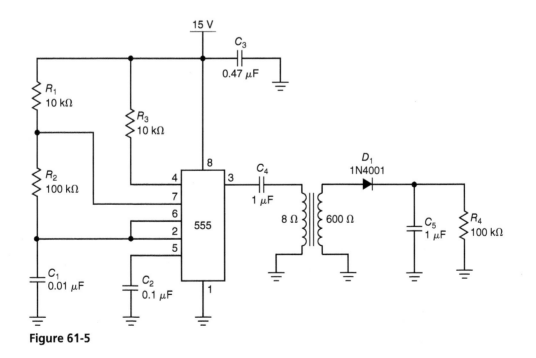

Figure 61-5

361

Experiment 61

Lab Partner(s) _____

PARTS USED Nominal Value	Measured Value	PARTS USED Nominal Value	Measured Value	PARTS USED Nominal Value	Measured Value
10 Ω		47 Ω		0.1 μF	
22 Ω		68 Ω		0.22 μF	
33 Ω		150 Ω			

TABLE 61-1. VOLTAGE REGULATOR

V_{in}	Estimated V_{out}	Measured V_{out}	
		Multisim	Actual
1 V			
5 V			
10 V			
11 V			
12 V			
13 V			
14 V			
15 V			

TABLE 61-2. RIPPLE REJECTION

	Calculated	Measured	
		Multisim	Actual
V_{rip} (min. rejection)			
V_{rip} (typical rejection)			

TABLE 61-3. VOLTAGE AND CURRENT REGULATION

	Calculated		Measured			
			Multisim		Actual	
R_2	V_{out}	I_{out}	V_{out}	I_{out}	V_{out}	I_{out}
10 Ω						
22 Ω						
33 Ω						
47 Ω						
68 Ω						

TABLE 61-4. TROUBLESHOOTING

Trouble	Estimated V_{out}	Measured V_{out}	
		Multisim	Actual
R_1 shorted			
R_1 open			
R_2 shorted			
R_2 open			

TABLE 61-5. CRITICAL THINKING: $R_2 = $ _____

	Multisim	% Error	Actual	% Error
V_{out}				

Questions for Experiment 61

1. When the input voltage of Fig. 61-1 is greater than 10 V, the output voltage is approximately: ( )
   (a) constant;   **(b)** 5 V;   **(c)** regulated;   **(d)** all of the foregoing.
2. In Table 61-2, the typical output ripple is approximately: ( )
   (a) 0.2 mV;   **(b)** 1 mV;   **(c)** 1.59 mV;   **(d)** 10 mV.
3. If I_Q is 8 mA in Fig. 61-3, the calculated I_{out} is approximately: ( )
   (a) 8 mA;   **(b)** 23.3 mA;   **(c)** 41.3 mA;   **(d)** 100 mA.
4. When R_2 is 68 Ω in Fig. 61-3, the calculated V_{out} is approximately: ( )
   (a) 5.43 V;   **(b)** 6.43 V;   **(c)** 7.04 V;   **(d)** 7.94 V.
5. The measured current in Table 61-3 indicates that the regulator circuit can function as a(n):   ( )
   (a) current source;   **(b)** voltage source;   **(c)** ripple generator;
   (d) amplifier.
6. Briefly explain why the data sheet of an LM7805 indicates that the input voltage must be at least 7 V.

7. Why are bypass capacitors used with an IC regulator?

TROUBLESHOOTING

8. What was the measured output voltage when R_1 was open? What caused this value of voltage?

CRITICAL THINKING

9. What value was used for R_2 to get an output of 9 V? How was this value determined?

10. Optional. Instructor's question.

Appendix

Main Parts and Equipment

RESISTORS (ALL ½ W UNLESS OTHERWISE SPECIFIED)

Quantity	Description		Quantity	Description
1	10 Ω		4	39 kΩ
2	22 Ω		2	47 kΩ
1	33 Ω		1	68 kΩ
1	47 Ω		2	100 kΩ
1	51 Ω		2	220 kΩ
2	68 Ω		1	270 kΩ
2	100 Ω		1	293 kΩ
1	100 Ω, 1 W		1	330 kΩ
2	150 Ω		1	470 kΩ
1	180 Ω		1	680 kΩ
2	220 Ω		2	1 MΩ
2	270 Ω			
2	270 Ω, 1 W			
8	300 Ω			
3	330 Ω			
4	470 Ω			
1	470 Ω, 1 W			
1	560 Ω			
2	680 Ω			
2	820 Ω			
1	910 Ω			
4	1 kΩ			
1	1.1 kΩ			
1	1.2 kΩ			
2	1.5 kΩ			
1	1.8 kΩ			
2	2 kΩ			
2	2.2 kΩ			
1	2.7 kΩ			
1	3.3 kΩ			
1	3.6 kΩ			
3	3.9 kΩ			
2	4.7 kΩ			
2	6.8 kΩ			
1	8.2 kΩ			
3	10 kΩ			
2	15 kΩ			
1	18 kΩ			
2	22 kΩ			
1	27 kΩ			
3	33 kΩ			

POTENTIOMETERS

Quantity	Description
1	1 kΩ
1	5 kΩ
2	10 kΩ
1	50 kΩ

CAPACITORS

Quantity	Description
2	100 pF
2	220 pF
1	470 pF
2	0.001 μF
3	0.002 μF
1	0.0033 μF
2	0.0047 μF
2	0.01 μF
1	0.02 μF
2	0.022 μF
2	0.033 μF
2	0.047 μF
1	0.068 μF
2	0.1 μF
1	0.22 μF
1	0.33 μF

CAPACITORS (cont.)

Quantity	Description
4	0.47 μF
3	1 μF
2	10 μF
1	33 μF
2	47 μF
1	100 μF
2	470 μF

DIODES

Quantity	Description
3	1N5234 (6.2-V zener)
1	1N5239 (9-V zener)
4	1N4001 (rectifier)
3	1N4148 (small-signal)
4	1N914 (small-signal)
1	L53RD (red LED)
1	L53GD (green LED)
1	2N6505 (SCR)

TRANSISTORS

Quantity	Description
1	2N3055 (power *npn*)
3	2N3904 (small-signal *npn*)
3	2N3906 (small-signal *pnp*)
1	IRF510 (power FET)
3	MPF102 (*n*-channel JFET)

INTEGRATED CIRCUITS

Quantity	Description
1	LM318C
3	LM741C
1	LM7805
1	LM7809
2	NE555

A Parts Kit, containing all of the listed Main Parts is available from:

APACO Electronics
6433 Pinecastle Blvd
Suite #2
Orlando, FL 32809
Phone: 800-261-3163
Fax: 321-319-0521
Email address: apcjr@att.net
Kit Number: #EP-8

MISCELLANEOUS PARTS

Quantity	Description
1	Inductor
1	TIL312 (7-segment)
2	SPST switch
1	small 8 Ω speaker

OPTIONAL PARTS FOR APPLICATIONS

MCM Electronics
650 Congress Park Drive
Centerville, OH 45459
phone (888) 235-4692
fax (800) 765-6960
http://www.mcmelectronics.com

Quantity	Description
1	Power Transformer 115vac To 12.6v - 2A CT MCM Part #: 287-1412
1	DC Motor - 11500 RPM 12VDC 160mA MCM Part #: 28-12810
1	Piezo Electronic Buzzer - 35mm MCM Part #: 28-740
1	7 Segment Display MCM Part #: 28-17828
1	Optocoupler MCM Part#: 107-4N25VM
1	Microphone Element - PC Board Mount MCM Part#: 35-4955

Digi-Key
701 Brooks Avenue South
Thief River Falls, MN 56701
Phone (800) 344-4539
Fax (218) 681-3380
http://www.digikey.com

Quantity	Description
1	Photoresistor Digikey Part #: PDV-P9008-ND